RIDDLES IN STONE

RIDDLES IN STONE

CONTROVERSIES, THEORIES AND MYTHS ABOUT SOUTHERN AFRICA'S GEOLOGICAL PAST

HUGH EALES

WITS UNIVERSITY PRESS

Wits University Press
1 Jan Smuts Avenue
Johannesburg
South Africa
http://witspress.ac.za

First published 2007.

ISBN 1-86814-447-X (10-digit ISBN)
ISBN 978-1-86814-447-1 (13-digit ISBN)

Edited by Andrea Nattrass
Cover design by Hybridesign
Cover photograph by Marion Bamford
Layout and design by Hybridesign
Printed and bound by Creda Communications

CONTENTS

PART 3

FOSSILS AND THE BEGINNINGS OF LIFE

PART 4

ROCKS, MINERALS AND OUR UNMATCHED ORE DEPOSITS

Preface

For several hundred years, geological science has been a battlefield of conflicting ideas. These controversies were not simply minor disagreements here and there, but included a great many clashes of interpretation that grew rapidly into acrimonious and protracted exchanges. The battles were fought within massive grey areas which some would stake their reputations on being white where others saw only black (or red!). Some saw the Earth as growing larger all the time, while others saw it shrinking. Early estimates of the Earth's age as around 6 000 years have been progressively revised upwards to the current 4 600 million years – a figure that is 766 000 times greater than the initial estimate. Fifty angry years were to pass before those who claimed that the continents wander on the surface of the Earth were able to silence those who obstinately maintained that this was nonsense – a mere 'fairy tale'. Minerals were once invested with magical properties, and some species were believed to be capable of reproducing themselves in the Earth by sexual union, before science showed them to be crystalline solids of atoms chemically bonded together as molecules. Earth science and religious doctrines have fought bitterly over evolution as opposed to Creation, over the Biblical Flood as opposed by geological evidence, and even over which of the Sun and the Earth revolves around the other. Diametrically opposed views have all too often allowed for little compromise or middle ground.

My original title for this book was *Some Good and Some Bad Calls in Earth Science.* My particular focus was on South Africa, as a collection of some of the ideas that lie submerged and largely forgotten in the literature of geology, but colleagues suggested that this gave an impression of frivolity. In defence of my original title, which I still prefer, 'Bad Calls' are in my lexicon predictions or statements that were made, perhaps in good faith, but which have turned out to be wrong, controversial or merely mischievous. As an illustration of what I would regard as a 'Bad Call' drawn from the field of history, the name 'Holy Roman Empire', slated by Voltaire as neither Holy, nor Roman, nor an Empire, would be a good example. The prediction in 1890 that the streets of London would be six feet deep in manure by 1920 because of all the horse-drawn traffic would be another.

This is not a textbook or reference book, nor does it pretend to give answers to the questions raised. It falls, perhaps, somewhere between what an older generation would recognise as a 'Commonplace Book' – a collation of the thoughts and beliefs of others which, for one reason or another, are perceived as memorable or significant – and a history of the development of some ideas in Earth science. My aim is to show how complex the planet on which we live is, and how few 'final answers' can be accepted with confidence. It is at the same time an attempt, however inadequate, to show that the evolution of geology as a science has produced a literature that should captivate any addict of 'popular science', without leaving him or her irretrievably lost in the technical language of the specialist. In the writing of it, I have used a broad brush rather than a fine nib, to exclude much tedious detail. The claim of physicists that 'physics is phun' is countered here with 'geology is global'. This is as much a 'bedside book' for geologists, biologists and geographers as it is a volume written in non-technical language for the miner, investor, layperson, or student. Given that the emphasis is as much on entertainment as on enlightenment, it is a pot-pourri of scientific controversy and myths waiting to be debunked.

In the emergence of Earth science as a reputable science, rather than a field for mere common-room speculation, we learn that today's dogma may be the heresy of tomorrow. If there is one thing the evolution of Earth science has taught us, it is that consensus amongst the majority, at any given time, is not proof of anything. Any science that calls for the interpretation of events that occurred hundreds of thousands to billions of years ago, and commonly at depths below the surface which we cannot ever hope to reach, must rely heavily on laws and observations based on the phenomena we see around us today. Therein lies the rub. There is no guarantee that things in the past – the force of gravity, the composition of the atmosphere, the climates of the world, the length of the day, the geographic position of the poles, the size of our Earth, or the map of the globe – were then as they are today. I leave it to you to decide whether Hutton's and Lyell's doctrine of Uniformitarianism (that the present gives us the key to what happened in the past) is to be trusted or not.

Most, but not all, of the pages that follow relate to South African geology, or concepts that originated elsewhere but became the foundations on which our interpretation of the geological history of this country was based. Some items come from my long-past days as a student; some from years as a mine and field geologist in Tanzania, Zimbabwe and South Africa; some from sabbaticals and conferences in Britain, Europe and the United States of America (USA); and the rest from standing at the front of a university classroom. It is quite likely that my contribution will be slated by colleagues for not also including 'this' or 'that' from the past. In pre-emptive reply, I can only plead that the problem I faced lay in what to include and what to leave out of a relatively short book.

When I was an undergraduate, there was a story about a history student who made a bizarre collection of 'versatile facts' to be learned before each examination – facts that could be worked into any essay, no matter what the topic, but which ought surely to impress the examiner by displaying a convincing breadth of general knowledge. For example, he learned the date on which the first camel was shipped into Ireland, and this he considered could be used to great effect in offerings such as: 'This piece of legislation relating to industrial practices was debated in the British parliament in the year 18~~, curiously enough, thirteen years before the first camel was introduced into Ireland.' This kind of approach to examinations cannot be recommended as sound strategy for student readers, but the pages that follow will at least serve the purpose of giving a background to some of the wayward ideas that have surfaced during the development of Earth science.

Hugh V. Eales
Grahamstown
May 2007

ACKNOWLEDGEMENTS

In the assembling of any manuscript, a writer is apt to trade on the goodwill of colleagues, thereby making himself something of a pest. For their good-natured forbearance, my grateful thanks go to each of the following for filling in some gaps, or for their comments on draft sections of my text.

At Rhodes University: Dr Emese Bordy (geology); Dr Steffen Büttner (geology); Prof. Chris de Wet (anthropology); Mrs Ashley Goddard (computing); Dr Thelma Henderson (Centre for Social Studies); Prof. Roger Jacob (geology); Prof. Julian Marsh (geology); Prof. John Moore (geology); Dr Rose Prevec (geology); Dr Steve Prevec (geology); Prof. Douglas Rivett (chemistry); Mr Mike Skinner (geology); Mrs Sue van der Riet (university library) and Prof. Marius Vermaak (philosophy).

From further afield: Mrs Jenny Bennie (historian at the Port Elizabeth Museum); Mr Leopold Bosch (executive manager of the Geological Society of South Africa); Mr Tim Broderick (geological consultant, Harare); Dr Billy de Klerk (Albany Museum); the late Prof. David Hammond-Tooke (anthropology, University of the Witwatersrand); Mr Athol Hockey (my son-in-law in New Zealand); Dr G. Levin (archivist of the Geological Society of South Africa); Mrs D. Nash (historian at the Port Elizabeth Museum); Dr Malcolm Roberts (Council for Geoscience, Port Elizabeth); Prof. Russell Shone (University of Port Elizabeth); Mrs Sue van Coppenhagen, (my daughter in Cork, Ireland for the watercolour of the ancient Zimbabwean rock crusher); Dr Mauritz van der Merwe (independent geological consultant); and Mrs Adri van Heerden (Council for Geoscience, Pretoria).

Particularly warm thanks go to Prof. Roger Jacob, for reading the manuscript from cover to cover, and offering comments and suggestions. Likewise, Andrea Nattrass and Estelle Jobson of Wits University Press proved themselves invaluable in the task of editing and arranging the text. Thank you too to Karen Lilje of Hybridesign for her layout, cover design and illustration, Lee Smith for proofreading and Margie Ramsay for indexing.

Grateful acknowledgement is made to the Geological Society of South Africa for their permission to scan and reproduce illustrations from early issues of their journal, *Transactions*, and to reprint the short article on gemstones written by my daughter, Sue. Similar acknowledgment goes to the Council for Geoscience in Pretoria for permission to reproduce several photographs from my book on the Bushveld Complex, published by them as *Popular Geoscience Series 2*.

The support given by the Dean of Research of Rhodes University, since my retirement from teaching, is recorded with sincere appreciation.

Above all, grateful thanks go to my long-suffering wife, Anne, for her help, for her advice while reading the text through the eyes of a layperson, and for tolerating a permanent snowstorm of papers in the home.

The publishers and I wish to thank Marion Bamford for her photo used on the cover and Byron Kozakiewiez of Beith Digital for assistance with scanning the images. We wish too to thank all those institutions, from whom extracts in their archives and publications have been reproduced and those individuals and institutions who have granted us permission to reproduce their images. Acknowledgement has been made accordingly at each instance.

A SIMPLIFIED INTERNATIONAL GEOLOGICAL TIMESCALE

ERA	PERIOD	EPOCH	AGE AT START (millions of years)
CENOZOIC	**Quaternary**	Recent	0.008
(Recent life)		Pleistocene	1.8
	Tertiary	Pliocene	5.3
		Miocene	23.8
		Oligocene	33.7
		Eocene	55.5
		Palaeocene	65
MESOZOIC	**Cretaceous**		145
(Middle life)	**Jurassic**		213
	Triassic		248
PALAEOZOIC	**Permian**		286
(Ancient life)	**Carboniferous**		360
	Devonian		410
	Silurian		440
	Ordovician		505
	Cambrian		544
PROTEROZOIC			2 500
ARCHAEAN			3 800
HADEAN			4 500
AGE OF THE EARTH			4 600

These ages should not be regarded as cast in concrete, as they are subject to minor adjustment from time to time. The dates here are those adopted by the United States (US) Geological Survey in 2002.

PART ONE

Evolution of Ideas About our Earth

1
Introduction: From Pythagoras to Plate Tectonics

'. . . a sketch of the progress of Geology is the history of a constant and violent struggle between new opinions and ancient doctrines . . . supposed to rest on scriptural authority.'
Sir Charles Lyell (1834: 43)

'Students of Nature, while eagerly pressing forward in the search after her secrets, are apt to keep the eye too constantly fixed on the way that has to be travelled, and to lose sight and remembrance of the paths already trodden.'
Sir Archibald Geikie (1905: 1), Director-General of Geological Survey of Great Britain

If you were to ask that shadowy and elusive character, 'the average person', what he or she understands by the word 'geology', the odds are you would get something close to: 'It's a lot of stuff about rocks.' There would be nothing intrinsically wrong with this response, but it would be woefully inadequate, conveying nothing about the true thrust of the science. Earth science has, particularly within the last 100 years, spread so wide as to spill over into the fields of chemistry, physics, biology, astronomy, ecology, climatology and engineering, and from there into economics, agriculture, and even politics. Perhaps a better reply would have been that geology is 'a reaching into the past in order to understand better the present and, possibly, the future'.

Earth science addresses questions that are fundamental. What is the meaning of those intriguing objects known as fossils? Was life created in the way that religious texts tell us, or did it evolve spontaneously? If so, when and where? How old is our Earth? Will there be an end to life as we know it today, or even to Earth itself? How long will the natural resources on which our civilisations have been built last – resources such as oil and coal or other fuels, metals, fertile soils, and even clean air and water? Does the unequal distribution of these natural resources between different nations hold the seeds of endless strife in the future?

We could not even begin to ask these questions without first looking back to trace the beginnings of our study of Earth. Here we find a story of painfully slow disentanglement of fact from myths and legends stretching back more than four millennia. It is a story of conflict with religious doctrines of both Christian and other theologies. And, above all, it is a story that called for the rejection of exuberant imagination that passed for science in earlier days.

THE GREEK AND ROMAN CONTRIBUTION

We are still only at the beginning of an understanding of the planet on which we live, and most of what we do understand about it has been learned in the last 200 years. It seems probable that we shall never learn precisely how much was known about the surface of our Earth in the times of the ancient civilisations of Mesopotamia, Egypt, Greece and Rome. The burning, by Arab invaders in 642 AD, of the 700 000 books lodged in the great library of Alexandria in Egypt, was an act of senseless vandalism that robbed the world of a vast store of collected knowledge about the 'Ancient World'. Priceless manuscripts were simply bundled together and burned to heat bath waters. Thus, rightly or wrongly, our surviving literature attributes the proposition that the 'Earth is round' to Pythagoras, the Greek mathematician of the sixth century BC, and the assertion that the Earth revolves around the Sun, rather than vice versa as Pythagoras would have it, to Aristarchus in the third century BC. In that same century, an estimate of our globe's circumference was made by the Alexandrian scholar, Eratosthenes (276–196 BC), using the length of shadows cast by the Sun at Alexandria and Assuan in Egypt. Even allowing for a fortuitous mutual cancelling out of several errors, the figure he derived for this (around 25 000 miles or 40 235 km) was astonishingly close to present-day estimates (24 902 miles or 40 077 km at the Equator).

Given a start such as this in ancient times, we might have anticipated rapid growth of the geological sciences in the succeeding millennia, but this was not to be. We find that, in large measure, it was theological dogma and the repressive philosophies adopted in Europe in the Christian era, that effectively put the lid on further meaningful progress for the next 1 700 years.

While we might have hoped that the 37 volumes of Pliny the Elder's (23–79 AD) *Natural History* would give a picture of the status of Earth science in the time of the Roman Empire, the critical review of his work by Healy (1986: 112) warns us not to expect too much – 'the Greeks and Romans had only a limited knowledge of geology'. Pliny offered some thoughts on earthquakes, volcanoes, and the properties and uses of the few minerals and ores that were known to him, but he showed little real insight. His assertion that rock crystal (the common mineral, quartz) 'is a kind of ice . . . [and] the inevitable conclusion is that rock-crystal is formed as moisture from the sky falling as pure snow' (Healy, 1986: 114–15) does not encourage us to take him seriously. Myths about diamonds, galena and limestone regenerating themselves in the Earth after their removal by miners seem to have been swallowed whole by him.

Figure 1: Natural crystals of the mineral quartz (silicon dioxide or SiO_2), one of the most common substances on Earth, which the Roman, Pliny the Elder, had thought was 'a kind of ice', believing that it was formed from hardened snow. This specimen is about 7 centimetres across.

The most influential English geologist of the nineteenth century, Sir Charles Lyell, had had little regard for him: 'Pliny had no theoretical opinion of his own concerning changes in the Earth's surface . . . he restricted himself to the task of a compiler, without reasoning on the facts stated by him . . .' (Lyell, 1834: 26). As regards 'Plinian Astronomy', Pedersen (1986: 173–84) judges that 'he did not really know what he was speaking about'. Most contemporary literature depicts Pliny as merely a compulsive scribbler, but Geikie (1905: 27) judges him rather more generously. Recalling his demise in the devastating eruption that destroyed Pompeii (in 79 AD), Geikie grants him the dubious distinction of being 'the first definitely recorded martyr to the cause of geological science'.

Regrettably, the enlightened ideas of Pythagoras and Aristarchus had made little real impact, for it was the teachings of Aristotle (384–322 BC), and then Claudius Ptolemy of Alexandria in Egypt in the second century AD that were to hold centre stage until the close of the fifteenth century. Both had taught that the Earth was the centre of the universe – a doctrine eminently acceptable to the Christian Church, which adjudged it a philosophy that was closest to the literal text of the Bible. Such was the power held by the Church, and the horrific punishments meted out by its Holy Inquisition, that any departure from this dogma would raise the spectre of trial, excommunication, hideous torture and execution. As regards Aristotle's own contribution to geology, his belief that subterranean winds imprisoned in the Earth were the cause of both earthquakes and volcanic eruptions does not greatly encourage further pursuit of his ideas in Earth science. Perhaps here was the origin of the view, whimsically expressed by one writer some 2 000 years later, that volcanoes are due to 'flatulence within the Earth's crust'? In reality, Aristotle has earned a low rating amongst scientists. In a paper delivered before the South African Association for the Advancement of Science, the Professor of mathematics at Natal University College, W.N. Roseveare (1917: 34), was particularly scathing: '. . . for want of accumulated facts and from a wish to produce results, he [Aristotle] generalized too rapidly, so bequeathing . . . a body of unscientific axioms which became dogmas, and thereby clogged the wheels of Science . . .'

BITTER CONFLICT WITH RELIGIOUS DOCTRINES

Under the strictures imposed by the Church, progress in the scientific study of planet Earth remained largely suppressed until the time of

Nicolaus Copernicus (1473–1543) – who dared to use observation rather than mere dogma or abstruse logic to construct a model of our solar system, with the planets revolving around the Sun. But, under the very real threat of the charge of heresy, he was understandably hesitant about broadcasting such ideas too boldly – and was in any event forced to recant his conclusions.

The determination of theologians to stifle 'heretical ideas' can even be seen in the standing regulations of the University of Louvain in Belgium in the stipulation that 'No one may be allowed to reject as heretical an opinion of Aristotle . . .' (see Crane, 2003: 47). Academic freedom was simply not on the agenda, and those who stepped out of line put their lives at risk. The map-making genius, Gerard Kremer (1512–94, better known as Mercator) became ensnared in this net, and found himself imprisoned at the instigation of the theologians of Louvain for corresponding with Protestant Lutherans, and for expressing doubts about the infallibility of the Aristotelian philosophers. He was lucky to escape execution, but the Antwerp printer, Liesvelt, who had published maps to accompany the Lutheran Bible, was decidedly less lucky – he was beheaded by the Inquisition in 1544 (Crane, 2003: 94, 162). Just how limited the advances in accurate mapping of the world had been during the preceding millennium, under such constraints, are illustrated by Mercator having devoted a substantial part of his life to recreating the 27 maps of the known world prepared 1 400 years earlier by Ptolemy – a task that he finally completed on publication of the work in 1578.

It was not the Christian Church alone that had stifled the development of the embryonic science. In the tenth century, for example, Omar 'The Learned' had compared the present coastline of Asia with ancient Persian and Indian maps dating back two millennia, to conclude that sea levels had dropped with the passage of time. Lyell (1834: 30) tells us that the scholars of the Koran had demanded that he recant his errors. Rather than do this, Omar had gone into voluntary banishment from Samarkand to escape persecution.

THE FIRST STIRRINGS OF SCIENTIFIC IDEAS

Georgius Agricola (1494–1555) was a sixteenth-century German miner, engineer and metallurgist. In his *De Re Metallica* (1556) we find the first glimmerings of a scientific approach to geological science in western Europe, and to the recovery and treatment of valuable ores.

Without the translation of his work from Latin by Hoover and Hoover (1912), Agricola would probably have remained a shadowy and obscure figure, unrecognised as one of the first to attempt a classification of minerals and a rational explanation for the origin of ore deposits.

Perhaps the real tribute for initiating a new start should go to someone who came later – Nicolas Steno (1638–86), the Dane who settled in Florence – for his perceptive contributions to the study of fossils, the geometry of natural crystals, the laws of superposition of rock strata, and physical geology. His dissection of a modern shark, which showed the teeth and bones to be identical to those of fossils found in Tuscany, was a massive step forward. Nevertheless, bound by theological doctrine, Steno had still clung to the custom of equating fossiliferous rocks with the mythical Biblical Flood.

Nothing, perhaps, better illustrates the obduracy of those who trod the corridors of power in Europe than the persecution of the wonderfully gifted Italian, Galileo Galilei (1564–1642), mathematician, engineer, astronomer and physicist. A graduate of the Universities of Pisa and Padua, he was already in trouble before the age of 30 for exposing fundamental mistakes in Aristotle's physics. In 1604, his experiment of simultaneously dropping a cannon ball of 200 pounds weight and a small musket ball from some height, and contrasting the speeds at which they fell, was a slap in the face for Aristotle, who had taught that falling bodies drop at a speed proportional to their weight (Bolles, 2000: 415–19).

When Galileo went so far as to defend the ideas of Copernicus, he trod dangerous ground: the work of Copernicus was declared heretical, and Galileo's own beliefs then condemned. He was forbidden in 1616 to publish in future any ideas that contradicted the belief that the Sun revolved around the Earth. Matters really came to a head in 1632 with the release of his new book, *The Two Chief World Systems*, which created such a storm that he was again dragged before the Inquisition in 1633 (Sobel, 1999). Despite proving that he had obeyed the decree of Pope Leo X in 1515 that all manuscripts should be submitted to the local bishop for approval and censoring before permission to publish would be granted, his book was condemned as further heresy. It remained on the list of banned works for the following 200 years. Galileo might perhaps have taken solace from the book then appearing on the black market at twelve times the original price, and being translated into other languages, but this did not persuade Rome to revoke his sentence of imprisonment. Although this was later reduced to house arrest and isolation from all other scientists, the Church still had the last word in forbidding the erection of a monument to him on his death in 1642. Only in 1818 were Galileo's writings removed from the list of forbidden works by Pius VII, who insisted that '. . . they should wipe off this scandal from the church' by repealing the edicts against both Copernicus and Galileo (Lyell, 1834: 99).

John Milton's *Paradise Lost* of 1674 is fascinating in the way it reveals that it was still prudent to 'play it safe' in the late seventeenth century when it came to interpreting matters such as Creation, and the status of planet Earth in the universe. In his Book VIII there is stern injunction against speculating about which body revolves around the other:

To ask or search I blame thee not . . .
. . . whether Heav'n move or Earth,
Imports not . . . (lines 66–71)

Whether the Sun predominant in Heav'n
Rise on the Earth, or Earth rise on the Sun . . .
Solicit not thy thoughts with matters hid,
Leave them to God above . . . (lines 160–68).

Given the prevailing climate of these times, it is not surprising that geology was a late arrival in the theatre of reputable science, lagging well behind mathematics, navigation, instrument-making, physics, and the science of chemistry emerging from the fog of alchemy and the search for the Philosopher's Stone to transform base metals into gold. Virtually every aspect of Earth science held the seeds of dangerous controversy – the concept of Creation, the mathematics of the solar system, the age of the Earth, interpretation of the legend of the Biblical Flood, evolution and extinction of living creatures, and the nature and origin of fossils. Nothing was tolerated that might contradict the doctrine that land and sea were separated on the third day of Creation, with animal life beginning only on the fifth day.

The point is driven home by Geikie (1905) devoting 41 pages in his *Founders of Geology* to recalling the evolution of Earth science during the time of the Greeks and Romans, but a mere 1.5 pages to the succeeding period extending to the sixteenth century, and then close on 450 pages to the times following Leonardo da Vinci (1452–1519). As McGowan (2003: 2) puts it in *The Dragon Seekers*, up to the late eighteenth century: 'English theologians and clerics had no argument with those who tinkered with chemistry or physics, but intellectuals who questioned the biblical account of the Creation could expect the full and considerable weight of the church to bear down upon them.'

Amongst those 'tinkerers' whom we would not normally associate with the growth of Earth science we might include Robert Hooke (1635–1703), physicist and mathematician and originator of 'Hooke's Law' ('the stress is proportional to the stretch of a spring'). As recalled by Mather and Mason (1939: 27–32), Hooke ventured into fields beyond physics – the elevation of land by earthquakes, erosion of coasts by the sea, and the deposition of sand and mud by rivers such as the Nile, the Thames, and the Po. He had had no reservations about certain fossils being 'real shells of Fishes, which . . . we conclude to have been at first generated by the Plastick faculty of the Soul or Life-Principle of some animal'. These are hardly explicit phrases, but he left his readers in no doubt about his scorn for those who believed in the then popular notion of an imaginary influence of the stars or 'any Plastick faculty inherent in the Earth itself so form'd' in creating fossil remains.

TIMES WHEN A VIVID IMAGINATION WAS AS PERSUASIVE AS FACTS

While it has taken millennia to chart the major visible features of the Earth's surface, establishing what lies deep beneath it came relatively late, for it was only in the nineteenth and twentieth centuries that the tools for doing this were conceived and built. There were many fundamental questions to be answered before any acceptable model could be offered. How old is the Earth? Has it been expanding or contracting with time? Is its deep interior hot or cold, solid, liquid or gaseous? Have the continents drifted on the surface of the globe, like great ships ploughing through the oceans? It is the twists and turns that followed during the search for answers to these questions that the remaining pages of Part 1 of this book set out to explore.

If our knowledge of what lies deep beneath the surface of the Earth came only relatively late, a better understanding of the sea floor came even later. From the evidence of fossils and their association in some cases with layers of salt, it had been accepted as far back as times following the seventh century BC (by Xenophanes, Herodotus, Aristotle, Strabo and others) that much of what is now dry land had once been covered by the sea. Cities, monuments and temples submerged beneath the waves told them that the reverse was also true. In general, however, the contributions of the Greeks and Romans to marine geology were of little real value. As Geikie (1905: 40) put it, 'nothing . . . had yet been accumulated to which the name of science could be applied'.

The level of knowledge about the oceans in the sixteenth century, and the absurd notions that prevailed, are illustrated by Mercator's 1569 map of the Arctic regions depicting the North Pole symmetrically surrounded by four fictitious islands. A torrent of north-flowing sea water was believed to stream through the narrow straits between these islands, and then be 'absorbed into the bowels of the Earth' down what Crane (2003: 236) has aptly referred to as an imaginary cosmic plughole. The belief in a universal ocean covering the entire Earth in its earliest times was a recurring theme in the seventeenth and eighteenth centuries, but it faced the twin dilemmas of providing, first, a source for the immense volumes of water required, and then, equally tricky, a mechanism for subsequently getting rid of it.

The French physicist and naturalist, Leclerc, Comte de Buffon, had in 1749 looked for an answer in a fracturing of the Earth's crust to allow the waters to drain away into mysterious caverns in the interior. This was close to the ideas of the mathematician, Leibnitz, who had in 1680 pictured a burning, luminous Earth developing a cavernous crust into which had drained a universal ocean whose waters had condensed as the planet lost heat (Mather & Mason, 1939: 45). As it so happens, this was exactly the reverse of what Thomas Burnet had concluded, in 1681, to explain the Biblical Flood, contending that the heat of the Sun had cracked open the Earth's crust to release a vast amount of water imprisoned there. This extraordinary belief in the interior of the Earth being filled with water surfaced yet again in 1695, in the teachings of John Woodward at the University of Cambridge.

An illustration of the way in which a fertile imagination could earn as much respectful attention as careful observation is found in De Buffon's *Epochs of the History of the Earth* written in 1776 (Mather & Mason, 1939: 58–73). In this flight of fancy, the smooth progression from 'what could have happened' to 'what probably happened' and the graduation of this to 'what did happen' is breathtaking. Starting with the premise that the impact of a comet on the Sun ejected 'little suns from the great one', and that these incandescent bodies then cooled and solidified, De Buffon accounted for the planets. The time required for this cooling he calculated as 'about 2 936 years for the Earth . . . and 1 130 for Mars'. During this cooling of Earth 'to the point of being able to touch it', the water vapour that was expelled into the atmosphere condensed to form a universal ocean.

The Earth was then, De Buffon calculated, 35 000 years old. Fossil shells in the Alps and Pyrenees mountains were 'definite proof that the seas covered the continent of Europe to 1 500 fathoms [2 740 m] above the level of the present sea', and to 3 660 metres in America, depositing sands, clays, slates and coals. The animals that the fossils represent 'could be regarded as the first inhabitants of the globe' that appeared during the first 10 000 to 15 000 years after the seas were established. The oceans being still at boiling point would have explained the very different species that thrived in times past. Because 'elephants have existed at the same time in Siberia and Canada', this 'proves incontrovertibly the continuity of Asia or of Europe with America' in early times. It was only 'when the waters withdrew' (in a manner unspecified) that the first volcanoes became active, because 'at the instant the fire is started by the effervescence of the combustible . . . stones, it is immediately extinguished by the water which covers it'. Hence, De Buffon declared, all active volcanoes are found near to coasts.

Returning to his theme of the remains of elephants being found in America, Europe and Asia, De Buffon concluded that these continents must have become separated by the subsidence of land, like Atlantis, as 'reported by Plato'. Everything in this extraordinary invention was tied to his timetable – the Earth was 75 000 years old, and expected to last for another 93 000 years.

Another outrageous view was that of Abraham Gottlob Werner (1749–1817), Professor at the Freiberg Mining Academy in Saxony. He led the small community who would in later years become known as geologists further up a blind alley than, perhaps, any other man. As a compelling theorist and magnetic speaker, he had persuaded a generation of scientists that the entire globe had at one time been encircled by a universal ocean that was as deep as the highest mountains that tower above the sea today. From this ocean, his 'universal solvent', all the older rocks around the globe had been chemically precipitated onto the sea floor in a fixed sequence that would everywhere be the same. This was the doctrine of Neptunism.

The oldest rocks Werner termed 'Primitive', to include the crystalline granites, gneisses (granitic rocks with banded structure), slates and sheets of basalt (that is, lava). There followed 'Transition Formations', 'Secondary' or fossiliferous rocks, and then 'Alluvial' or 'Tertiary' rocks. The paradox of where this enormous mass of water disappeared to after precipitation of the Primitive rocks, or how it then reappeared to distribute the younger rocks, was brushed aside. To explain away active volcanoes, he insisted that they were produced by the melting of rocks from the heat of burning coal seams.

As Werner's star and his influence waned, confusion followed with the realisation that some granites and schists that he had grouped as 'Primitive' were actually younger than his 'Secondary' or fossiliferous rocks, and that his Transition Formations could in turn also be found amongst the youngest or Tertiary grouping. Werner's classification of rocks by their textures and mineral content, and the attempt to equate this with their age, simply made no sense, but it was to take many years to wipe the slate clean of his teachings.

The tearing apart of Werner's ideas by the one-time Director of geological surveys in Britain, Sir Archibald Geikie (1905: 201–40), must stand as one of the most blunt 'demolition jobs' in all the literature of geology. While acknowledging Werner's contributions to the development of mineralogy and mining, and the magnetic personality by which he had persuaded his listeners to his own point of view ('No teacher of geological science either before or since has approached Werner in the extent of his personal influence . . .') Geikie went on to sweep away the last lingering vestiges of Neptunism in the scathing passages: 'Never was a system devised in which theory was more rampant; theory, too, unsupported by observation, and, as we now know, utterly erroneous. From beginning to end . . . assumptions were made for which there was no ground . . .', and 'it would be difficult to cite from any other modern scientific treatise a series of consecutive sentences containing a larger number of dogmatic assertions, of which almost every one is contradicted by the most elementary facts of observation'. Geikie then hammered the final verbal nails into Werner's coffin with the words: 'His death in 1817 was felt by many . . . to bring a relief from the despotism which he had so long exercised.'

Another grand delusion was that of the Chevalier de Lamarck (1744–1829), one-time soldier, medical doctor, botanist and finally, a Professor of zoology and founder of invertebrate palaeontology in France. He had in all seriousness asserted in 1802 that, under the influence of the Sun and Moon, the waters of the oceans must inexorably be drawn ever westwards, thereby causing the east coasts of continents to be attacked by the sea, while new sediments collected on their sheltered west coasts. His thesis was that, in this manner, the oceans, and by inference the continents, had circled the Earth several times. Equally misguided was his belief that all rocks – even granites and metallic ores – represent the remains of decaying organic debris. In his eyes, calcareous rocks or limestones stemmed from the remains of animals, and fine-grained, clayey rocks from those of plants.

Popular literature had been no better in developing a scientific outlook in subsequent years. In *The Rime of the Ancient Mariner*, Samuel Taylor Coleridge's (1772–1834) tale of the haunted ship and the albatross, the superstitions of sailors had the polar regions presided over by malevolent spirits, while 'slimy things did crawl with legs upon the slimy sea'. The myths of the early days died hard. In 1882, American politician Ignatius Donnelly became a millionaire from the sales of his *Atlantis: The Antediluvian World,* a book about the vanished mythical continent of Atlantis. The great store of treasures that mythology had attributed to a civilisation that was said to have flourished there became to treasure-seekers what honey is to the badger. Some believed it to have sunk into the North Atlantic, and others, the Mediterranean. Even today, programmes discussing in all seriousness the descent into the depths of an entire continent during the lifespan of civilised man appear at intervals on television, despite the flat contradiction of basic principles of physics.

HUTTON'S QUIET REVOLUTION

Geological science may be regarded as having finally moved towards reality with the publication by the Scot, James Hutton (1726–97), chemist, lawyer, medical doctor and farmer of Edinburgh, of a 96-page essay in 1785, and his *Theory of the Earth* in 1795. Unfortunately, he wrote in such a convoluted and opaque style that the reading of it required real fortitude, and apparently not many bothered.*

* How many editors of scientific publications would today allow to appear in print the following paragraph, selected by Geikie (1905: 310) as an illustration of 'Hutton's peculiar style of composition'?

> If, in examining our land, we shall find a mass of matter which had been evidently formed originally in the ordinary manner of stratification, but which is now extremely distorted in its structure and displaced in its position, – which is also extremely consolidated in its mass and variously changed in its composition, – which, therefore, has the marks of its original or marine composition extremely obliterated, and many subsequent veins of melted mineral matter interjected, we should then have reason to suppose that here were masses of matter which, though not different in their origin from those that are gradually deposited at the bottom of the ocean, have been acted upon by subterranean heat and the expanding power, that is to say, have been changed in a greater degree by the operations of the mineral kingdom.

How many would be able to grasp his meaning at the first reading of such a sentence, or realise, without re-reading it, that this was his description of the process of metamorphism (that is, recrystallisation) of rocks?

It was Hutton's friend, John Playfair, who saved him from obscurity in his *Illustrations of the Huttonian Theory of the Earth*, which came out in 1802, after Hutton's death. Here, Werner's ideas were ridiculed, igneous rocks were recognised as molten material intrusive into or extrusive onto the Earth's crust, and of different ages, while Hutton's celebrated maxim – that the present provides the key to the past – became enshrined in the principle of Uniformitarianism.

Nevertheless, the influence that Werner had wielded lingered still, and it was only as late as 1865 that Charles Lyell, the leading English geologist of the day, commented that the belief that all schists and plutonic (deep-seated) rocks such as granites were of necessity older than sedimentary and volcanic rocks 'is nearly exploded'. Here, however, Lyell himself went too far – not only did he reject the view that all granitic rocks had originated 'before the creation of organic beings', but he went to the other extreme (1865: 718) by maintaining that 'we find it no easy task to point out a single mass of granite demonstrably more ancient than all the known fossiliferous deposits'. This is to be seen as a classic case of misinformation being used to refute a totally invalid concept.

Rather less well known was Hutton's close friend, Sir James Hall, whose experiments on the melting of rocks must stand as the most instructive conducted up to that time in this field. Intrigued by the results of an accident in a glass-works, where a large mass of molten glass had ended up as a useless mass of crystals instead of as transparent glass, Hall began his investigations (Geikie, 1905: 319). The furnace in an iron foundry became his laboratory, into which he thrust his crucibles and sealed gun barrels holding experimental charges. From this he determined in 1790 that it was the rate of cooling that determined whether glass or crystals were produced. This led on to his showing that melts of natural rocks such as basalt (a dark-coloured lava) behaved in precisely the same way. This should finally have laid Werner's Neptunist ideas to rest, for it vindicated Hutton's beliefs in the nature of igneous rocks, but Sir James did not have the charisma of a Werner. And then, who would accept that experiments in little crucibles could possibly solve one of the most hotly debated questions in Earth science?

By now it should be apparent that a great deal (perhaps most) of what was written in the name of Earth science before the late eighteenth century may be disregarded as being merely absurd – the expression of quaint opinions based on the most slender evidence, or on none at all. Nevertheless, it would be short-sighted to disregard the insight and boldness of some earlier Italian thinkers such as Leonardo da Vinci (1452–1519) and the writings of Fracastoro (1517), or the Frenchman, Palissy (1580) as recorded by Lyell (1834: 33, 38), all of whom had long ago recognised the true origin and significance of fossils, and the way sediments collected on the floors of the oceans. Had their arguments been accepted earlier, the fog that had held back a better understanding of our Earth would have lifted several centuries sooner. If natural scientists had not been fettered by religious doctrines, it would have been easier for them to come to terms with the facts that pointed to the great age of the Earth, and the gradual evolution of all living things.

THE FLOORS OF THE OCEANS

Prior to the Second World War of 1939–45, very little was known about the geology of the sea floor, apart from widely spaced measurements of its depth made by employing weighted steel wires lowered from ships, and the investigation of atoll islands fringed by coral reefs, stimulated by Charles Darwin's theory of their origin – that slow but prolonged subsidence of volcanic peaks beneath the waves was accompanied by continued upward growth of corals striving to remain near the surface, in the shallow waters lit by the Sun. The development of echo-sounding instruments and magnetometers during the Second World War revolutionised all of this, and marked the start of a sustained programme of detailed mapping of the ocean floors. Maps of the sea-floor topography showed volcanic seamounts and flat-topped edifices (guyots) reaching from great depths towards the surface, deep trenches, mid-ocean ridges with steep-sided clefts along their axes forming the longest 'mountain chain' on Earth, and great fractures extending for hundreds of kilometres.

The recovery of cores led to the dating of sediments on the sea floor, and reconstruction of the depths and temperatures of water in the distant past. Ship-based seismic surveys provided data on the nature and thickness of sea-floor sediments. Rachel Carson's essay, 'The Long Snowfall' (1951: 439–46), painted a vivid picture of this early phase, when she wrote: 'the reading of the story contained in the sediments has only begun'. There were hopes that they would eventually reveal the history of the Earth back to the remotest of times, by preserving the evidence in the deep-sea clays, the fine dusts carried by winds or spread by volcanic eruptions, the iceberg-drifted debris, and the delicate shells of *Globigerina* and *Radiolarian* oozes that had drifted to the bottom. This was, however, the time before the process of sea-floor destruction and renewal became recognised, and the realisation that the ocean floors are much younger than our ancient continents.

It was only with the subsequent measurement of heat flow and unscrambling of the magnetic signatures of sea-floor rocks that continental drift became an acceptable hypothesis that was refined through the years 1940–70 into what must stand as the great triumph of the twentieth century – the integration of geology, geophysics, geochemistry, geochronology, oceanography and palaeontology into the unifying theory of Plate Tectonics.

SOUTHERN AFRICA EMERGES INTO THE LIMELIGHT

In one sense, it was perhaps fortunate that the piecing together of the geology of southern Africa began relatively late. The slate had been more or less swept clean of the implausible propositions of Aristotle, De Buffon, Werner and Lamarck, with the appearance in print of the works of James Hutton and John Playfair at the beginning of the nineteenth century, Charles Lyell in its middle years, and Archibald Geikie towards its end. There was by now a clearer understanding of the different classes of rocks, and of the nature and value of fossil assemblages preserved within them. Relative ages of the fossiliferous rocks were reasonably well established, even if their absolute ages were not. Descriptive and determinative

Figure 2: Andrew Geddes Bain, known as the 'father of South African geology', had in his lifetime been a hunter, a saddlemaker, a journalist and a supervisor of prisoners sentenced to hard labour. Self-taught from having studied Lyell's *Principles of Geology* (1834), he went on to find the first dinosaur fossils in South Africa, the first vertebrates in the Karoo Sequence, and publish the first geological map of the Cape Colony. He was also as much a pioneer in the building of roads and mountain passes (including Michell's, Ecca and Katberg Passes) as he was in the young science of geology. (Portrait scanned with permission from the *Transactions of the Geological Society of South Africa*, 1897.)

A memorial to Andrew Bain stands a few kilometres outside Grahamstown, at the top of the Ecca Pass. The inscription reads:

Andrew Geddes Bain, road builder and geologist 1797–1864. Bain built the Queen's Road to Fort Beaufort via the Ecca Pass and the road through Pluto's Vale as military roads in 1837–45. His house was then near here on the Ecca Heights. He became interested in geology in 1837 during the construction of the Queen's Road, and in the area visible from this spot he worked out the stratigraphy of the Karroo System and discovered that fossil reptiles occur in it. Bain was the father of South African geology.

Historical Monuments Commission, 1964

mineralogy had by now evolved into an advanced science, and with the invention of the polarising microscope by William Nicol in 1829, and its application to the study of rocks by Henry Sorby of Edinburgh, a tool of immense power had dropped into the hands of the Earth scientists.

Serious attention was first drawn towards southern Africa with the publication of a geological map of its southern regions, and some account of its fossil reptiles, by Andrew Geddes Bain (1797–1864) in the *Transactions of the Geological Society of London* in 1856. In 1845[1], Bain and Dr W.G. Atherstone had dug up the first dinosaur bones discovered in South Africa in the Bushmans River Valley in the Eastern Cape. Although the specimens had merely gathered dust in the vaults of the British Museum for 31 years before the unpredictable and quarrelsome Sir Richard Owen made the first descriptions of them (De Klerk, 1995a, 1995b), they initiated the birth of palaeontology – the study of fossils – in this country.

In Grahamstown Atherstone made the first identification of a diamond found in this country. Following their discovery in 1867, geologists worldwide began to focus on rocks formed deep beneath the Earth's crust. The world's greatest goldfield was discovered on the Witwatersrand in 1886 and the subsequent dominance of South Africa in the production of diamonds and gold made this country 'a must' on the itinerary of leading overseas geologists with the means to travel.

It was the brilliance of Alex du Toit that brought southern Africa into the spotlight in another field. His first contribution to the evidence for continental drift in 1921 was years in advance of the times, so much so that 30 to 40 years would pass before the true value of his work was realised amongst most geologists in the northern hemisphere. In 1924 the discovery of platinum-group metals in the Bushveld Complex, the largest igneous intrusion of its type in the world, drew visits by scientists from all parts of the globe. Here was the world's greatest repository of not only the platinum-group metals, but of other industrial metals such as vanadium and chromium. In the meantime, unique fossils from the Karoo Sequence were becoming recognised as bridging the gap between reptiles and the more advanced mammals.

Thus was the stage set in this country for the young science to explode in a burst of new concepts that would emerge from research on its uniquely diverse rocks, ores and fossils.

Figure 3: Dr William Guybon Atherstone, MD, F.R.C.S., F.G.S., was the son of an 1817 settler in the Eastern Cape. Like his father, he became a medical practitioner, and was the first man in this country to use a general anaesthetic (ether) during surgery. As a close colleague of Bain, he is best known for his contributions in geology, particularly in palaeontology, and for having made the first positive identification of a diamond found in this country. In 1881 he became a member of the House of Assembly of the Cape Colony, and was in 1895 elected an Honorary Vice-President of the Geological Society of South Africa at its first meeting. He died, afflicted by blindness, at the age of 82 in 1898. (Portrait scanned with permission from *Transactions of the Geological Society of South Africa*, 1898.)

POLITICS, ECONOMICS AND THE TRANSVAAL'S RICHES: AN EXPLOSIVE MIX

Without some understanding of the events that took place in southern Africa between 1886 and 1902, it would be impossible to appreciate the degree to which the mineral wealth discovered in this country came to influence both national and international affairs. South Africa was to learn, the hard way, that the possession of treasures embedded in its rocks was not an unmixed blessing.

For reasons that were not entirely altruistic, the government in Britain was becoming increasingly involved in southern Africa. The British flag already flew over the Cape Colony, but after the discovery of rich diamond deposits at Kimberley, 'the ensuing diamond boom in the Cape (which had, by some smart map-reading, hastily incorporated Kimberley in its territory) transformed the economy of the region' (Pakenham, 1982: 18). With the subsequent discovery of the riches of the Witwatersrand, 'it was gold that had lured Rhodes and Beit in 1895 to try to re-annex the Transvaal to the British Empire – a scheme in which [Joseph] Chamberlain and the imperial government (so they claimed) had played no part' (Pakenham, 1982: 20). This disclaimer is, however, at odds with the sentiment expressed by Chamberlain that the Jameson Raid might 'postpone for ever the chance of creating a British Federation of South Africa, built up on the gold of the Transvaal' (Pakenham, 1982: 28).

Nowhere in Pakenham's meticulous documentation of the events that led to the Boer War (now commonly referred to as the South African War) of October 1899 to May 1902 is it suggested that the war was motivated by mere acquisitiveness. We must wonder, however, what would have been the sentiments of the Colonial Office, running on a total budget only recently increased to a mere £600 000, when the reserves of the Witwatersrand goldfields were conservatively estimated at £700 million (Pakenham, 1982: 27).* Pakenham follows the line that the High Commissioner for South Africa, Sir Alfred Milner, precipitated the war for political and personal reasons: 'His [Milner's] plan was to annex the Transvaal. He would rule it as a Crown Colony', and at the same time secure 'a place in history as big as the man who made the American Constitution'. Subsequently, Milner admitted, 'I precipitated the crisis . . .' (Pakenham, 1982: 64, 114).

The views of others on the cause of the Boer War were less magnanimous. In his biography of General J.C. Smuts, *Grey Steel*, Armstrong (1938: 82) relates that Paul Kruger, President of the Transvaal Republic, had vehemently protested, 'It is our country you want!' Armstrong also argues that before the discovery of gold, 'The English and Dutch had continued to quarrel, but had kept their quarrels as those of relatives', but 'with the finding of gold South Africa changed and woke from sleep into a vicious mood . . . the English Government suddenly realised that South Africa was valuable and made up its mind to control its gold' (1938: 39).

In the event, the golden harvest of the Transvaal led, ultimately, to what Pakenham describes as 'the longest . . . the costliest . . . the bloodiest . . . and the most humiliating war for Britain between 1815 and 1914' (1982: xv). Some 22 000 British, 25 000 Boer, and 12 000 African lives were sacrificed.

* We can only guess what Chamberlain's reaction would have been, had he been able to look into a crystal ball and learn that gold production between the years 1887 and 2002 alone would yield some 47 300 metric tons, or more than 1.5 billion troy ounces of gold (1 troy ounce = 31.1 grams). If we were to use a value of US$ 650 per troy ounce as the median gold price at the present time, the value of this would equate to some US$ 975 billion, or roughly £500 billion.

Somewhere between 20 000 and 28 000 Boer civilians – elderly men, women and children – died of epidemics in the concentration camps, their farms were burned, and their stock looted.

Such were the tragic consequences of the explosive brew that resulted from the mix of political aspirations, economic factors, and the discovery of geological riches.

It would be quite misleading to imply that Earth science in South Africa then progressed without constraints through the remaining years of the twentieth century. In 1922, E.T. Mellor had wryly commented in his presidential address to the South African Association for the Advancement of Science on '. . . the very meagre support given by the government to geological investigation . . .', while at the same time dropping a remark that would be as appropriate today as it was then: '. . . the Witwatersrand attracts to it the notice of the politician, the labour reformer and the tax gatherer'.

A caustic but otherwise delightful quote has been preserved by Harold Harger, the last surviving founder-member of the Geological Society of South Africa in 1945. He recalls a derogatory remark made in earlier days by C. Wilson-Moore, another founder-member, that our geological knowledge was 'patchy and imperfect' and that 'the sayings and writings of those who have gone before us . . . have given us a jumble from which we can but pick out a few of the pieces which lie contiguously. As an incohesive heap they are well-nigh useless' (Harger, 1945: i). This was harsh and unfair judgement indeed, for by 1929 remarkable progress had, in fact, been made. Approximately half of the area of the Union of South Africa had been mapped, the distribution of the major groups of rocks established, and *The Geology of the Union of South Africa* by Rogers, Hall, Wagner and Haughton (1929) released in Germany, while published works on the Bushveld Complex and the Witwatersrand Sequence were attracting the attention of geologists around the world.

Modern-day geology, in which the boundaries between it and physics, chemistry and the biological sciences were to become blurred, had 'arrived' in southern Africa.

2
How old is the Earth?

'Heaven and Earth . . . were made in the same instance of time . . . and man was created on the 26th of October, 4004 BC, at 9 o'clock in the morning.'
As calculated by Dr John Lightfoot, Vice-Chancellor of Cambridge University, shortly after Archbishop James Ussher's 1654 pronouncement that the Earth was approximately 6 000 years old

'Oh, how vast is the antiquity of our earth! and how small are the ideas of those who assign to the existence of this globe a duration of six thousand and some hundreds of years from its beginning to our own days!'
Chevalier de Lamarck (1744–1829) as quoted by Geikie (1905: 356)

Basing his calculations on a model of the Earth cooling from an initially molten state, Lord Kelvin gave its age in 1897 as within the range of 20 to 40 million years.
Holmes (1965: 347)

' . . . the formation of the ocean to the present time constitutes the greater part of the age of the earth . . . the age . . . is therefore probably between 1.3 x 10^9 and 3 x 10^9 years.'
Sir Harold Jeffreys (1929: 80), Cambridge mathematician

'Isotopic and petrographic work on ancient Western Australian zircons implies granites and hydrosphere were present on Earth 4 400 to 4 200 million years ago.'
Nutman et al. (2001: 93)

The extracts presented above are merely a few of the milestones dotted along the winding path that eventually led towards establishing the true age of the Earth (at around 4 600 million years). The calculations of Ussher and Lightfoot in the seventeenth century had been based on a literal interpretation of the text of the Bible. In the light of present-day knowledge, the age they assigned appears absurd, but it is interesting that this turned out to be close to the date generally accepted in the mid-twentieth century for the ending of the last ice age (8 700 years ago; Holmes, 1965: 350) or the climax in global warming 5 000 to 7 000 years ago when 'the British Isles and comparable regions elsewhere enjoyed a climate more like that of the Mediterranean of today', suggestive of the Garden of Eden (Holmes, 1965: 692).

Figure 4: Detail of frieze from an earlier ninth-century church incorporated into the west wall of thirteenth-century St Declan's Church in Ardmore, County Cork, Ireland, illustrating Biblical themes. Although badly weathered after some twelve centuries of exposure, the stone carving clearly represents Adam and Eve, the first man and woman according to Creationist doctrine, under the Tree of Knowledge in the Garden of Eden, as presented in Genesis 2.

There were others who took a more liberal view of the biblical account of Creation within six days as a purely allegorical representation of a much longer period preceding the Biblical Flood. In this way, prominent figures such as William Buckland (1784–1856), the first Professor of geology at Oxford University, and Georges Cuvier (1769–1832), the leading anatomist and palaeontologist in Europe, reconciled their religious convictions with what the young science was telling them – that the Earth was at least tens to hundreds of thousands of years old (McGowan, 2003: 8).

THE FIRST HINTS GIVEN BY A SCIENTIFIC APPROACH

The estimates of Lord Kelvin (the most famous physicist of his day), of a few tens of millions of years, had been based on a model of the Earth as a hot, molten body spinning in space cooling down to its present congenial state. The crust on which we live was seen merely as the rigid, outermost skin. Between 1863 and 1897 Kelvin had repeatedly revised his figure downwards from 400 million years, to 100 million, 50 million, and finally 24 million (Hallam, 1989: 122). Most geologists felt that this was far too short to allow for the evolution of the diversity of living species – Lyell had insisted on a minimum of 240 million years for this, while Charles Darwin had come in for criticism for having stipulated 300 million years.

Subsequent work with modern data has confirmed the essential precision of Kelvin's calculations, if based on the premises of his model, but there was a fatal flaw embedded in his reasoning – the assumption that the Earth itself did not generate heat from within. The later discovery, led by Lord Rutherford in 1904, that heat is released by the decay of radioactive elements such as uranium, thorium and potassium within its rocks wholly invalidated Kelvin's approach. If the Earth had cooled more slowly than he assumed, the age had to be much greater than he had calculated.

Without the advantage of the 'self-winding clock' – radioactivity – that was subsequently found to be embedded in the rocks themselves, geologists of the nineteenth and early twentieth centuries could do no more than fall back on estimates that relied, ultimately, on Hutton's and Lyell's principle of Uniformitarianism. The essence of this doctrine is that what is happening today can be extrapolated back into the past. Those who followed after Hutton then reasoned, by extension, that if the rate at which a process acting at the present time could be established, then its cumulative effect through geological time should give a date for the starting point. Although this approach yielded answers that we now know to have been wrong by more than two orders of magnitude, it was at least a step forward in that it relied on actual measurements and calculation rather than on biblical notions, philosophical debate or simple prejudice.

The essay by Professor W.J. Sollas (1905) of Oxford offers in his *Age of the Earth* an illuminating review of where matters stood as the nineteenth century yielded to the twentieth. Three different techniques tried up to that time were singled out by him. Sir George Darwin had calculated a date for the separation of the Moon from the Earth as not less than 56 to 57 million years ago, which had prompted Sollas to enthuse that 'Professor Darwin's story of the moon is certainly one of the most beautiful contributions ever made by astronomy to geology' (1905: 7). This effusive comment was most likely prompted by Darwin's figure being of the same order as his own estimate.

Professor John Joly in Dublin had followed a different tack, based on the assumption that all oceans of the world had started as fresh water, to which salt had slowly been added through geological time by leaching of the land by rain and groundwaters that were recycled back into the sea.* Amongst the many flaws in Joly's approach we would now recognise the enormous amount of salt locked up in evaporite deposits of the world,[2] and salt recycled back into the Earth's mantle by the process of subduction, whereby crustal rocks are thrust to great depths as the continents shift their positions on the surface of the globe.

When assessing Lord Kelvin's estimate of 20 to 40 million years, Sollas had denounced the lower figure as too low, arguing that the assumed rate of increase of temperature with depth into the Earth was flawed, and hence 'Lord Kelvin's estimate of the date . . . may be pushed back into the remoter past' (1905: 20). The real flaw in Kelvin's method – neglect of the heat produced by radioactive elements – was mentioned only in passing by Sollas (1905: 64) but, surprisingly, without emphasis on the fundamental implications of this.

Estimates of the Earth's age seemed at that time to have settled somewhere within the range from 20 million to 100 million years, and Sollas clearly believed that the figures simply needed a bit of tightening up, and that would be the end of the matter. The time it had taken for the sedimentary rocks of the Earth's crust to accumulate was the solution he opted for. Professor John Phillips at Oxford University, and Sir Archibald Geikie had earlier arrived at a figure of some 100 million years for this, but Sollas now made his own assessment of the total thickness of all sediments built up through geological time — 265 000 feet (81 km), including 183 000 feet (55 km) since

* Should we be surprised to read that an essay on this very topic had sprung from the fertile brain of Leonardo da Vinci nearly four centuries earlier (Mather & Mason, 1939: 1) under the title 'Why Water is Salt'? What Joly did was to quantify the approach: his measurements and calculations had indicated that the total mass of salt in the oceans today is 90 million times that being carried into it each year, which would equate to a figure of 90 million years for the age of the Earth. In Sollas's view, this was an overestimate, on the grounds that Joly's determined average of 5.73 parts per million of salt in the waters delivered to the sea was rather too low, and based on inadequate sampling. His judgement was that: 'A review of these facts before us seems to render some reduction in Dr. Joly's estimate imperative. A precise estimate is impossible, but . . . I should be inclined to take off some thirty or forty millions of years' (Sollas, 1905: 23). This decidedly cavalier and arbitrary reduction in age by some 40 per cent seems to be rather more subjective than scientific.

As Holmes (1927: 39) pointed out, so many assumptions have to be made that the method 'cannot at present give results of any value'. Apart from everything else, measurements would be skewed by the Earth at present exposing more land, with higher average elevation and more vigorous erosion than average values since at least the Cambrian Period. Furthermore, the emergence of humans has also led to deforestation, accelerated erosion, pollution, and release of acids into the atmosphere, all of which would lead to a more rapid release of salt.

the Cambrian Period.* To this Sollas applied a rather arbitrary average of *1 foot per century* (30.5 cm/100 years) as the rate of accumulation. The time elapsed thus became 26.5 million years, and that since the beginning of the Cambrian Period 18.3 million years. He did acknowledge that there was a cavea – that during Earth's juvenile stages the oceans might have dropped from boiling point to become frozen at a time when the Sun was giving out little heat, and the ice-bound Earth was thus incapable of transporting or depositing sediments. In his mind, this 'age' would therefore represent time elapsed since the heat of the Sun had warmed the Earth, rather than the age of the Earth itself.

* Arthur Holmes (1927) subsequently determined the full sedimentary column as around 530 000 feet (160 km), that is, double Sollas's estimate.

So, where did Sollas's errors lie? For a start, he had assumed that the thickness of crustal sedimentary rocks preserved would be proportional to the time elapsed, and that average rates of accumulation could then be determined to establish how long it had taken for this deposition to occur. Both assumptions were wrong. Vast intervals of time might pass with little or no deposition at all, when or where there is a shortage of debris available. In addition, complete loss of sequences already laid down has occurred many times in the past.[3] Furthermore, there is in reality no such thing as a meaningful average,[4] as this is tied to many different factors – environment of deposition, rock type, periods of mountain building, world climates, etc. Striking an average would be rather like determining, for example, an average height for all plants – the arithmetic is a simple matter, but the disparity in height between a lowly moss and the stately sequoia tree would render the average quite useless for all practical purposes.

Another fundamental mistake, about which Sollas could have suspected nothing, emerged as the Plate Tectonic Model was developed in the latter half of the twentieth century. As new facts were gathered in support of this model, it was established that the ocean floors are all much younger than the ancient continents – nowhere older than 200 million years – because the older deposits accumulated there are continuously displaced as new sea floor is created while the continents shift position on the Earth's surface.

It is intriguing to find, also, that there was a resounding contradiction embedded in Sollas's work. He had had little time for Hutton's doctrine of Uniformitarianism, dismissing it as '. . . a dogma in the narrowest sense of the word; unproved, incapable of proof . . .' (1905: 3). He had castigated Lyell for his belief in it, accusing him of having '... consistently refused to extend his gaze beyond the rocks at his feet, . . . thus [being] led to do serious injury to our science . . .' (1905: 3). Yet, the method selected by Sollas himself was firmly rooted in Uniformitarianist doctrine in pure form.

In spite of the fact that five different world leaders had come up with answers within the range 20 million to 100 million years for the age of the Earth, all were massive underestimates.

GLACIAL VARVES: COUNTING THE YEARS ONE BY ONE

The counting of layers in thick sequences of varved sediments (sediments deposited by rhythmic seasonal fluctuations) relies on an alternation of coarse and fine-grained, or darker and lighter sedimentary layers (see Plate 1). In regions with frigid climates, the melting of ice sheets or glaciers during the warmer months spreads coarser-grained and thicker layers on the floors of placid bodies of water when rivers run vigorously, and thinner, finer-grained layers during the frigid months when the flow of surface waters is greatly impeded. Each pair of layers therefore represents one year.

Dickin's (2002) review of the search for physical evidence of the Biblical Flood in Mesopotamia showed how varves may be used to calibrate carbon-14 ages back to 36000 BC, but this does not even enter the frame for measuring the age of the Earth. The method is also severely limited because it tells us nothing about the enormous intervals that might have elapsed between one glacial period and another. But, what it did show to the early investigators was that the duration of epochs during which varved sediments were laid down was very much greater than most previous estimates had implied. For example, the duration of that epoch known as the Eocene (determined in 1929 as around 22 million years in North America by varve counting; Holmes, 1965: 351) made it virtually impossible to squeeze all older fossil-bearing rock formations into what was then taken to be the span of time since the start of the Cambrian Period – a mere 28 million years according to Walcott, the Director of the US Geological Survey in 1893.

RADIOACTIVITY: THE GREAT BREAKTHROUGH

The great breakthrough came with the recognition of the significance of radioactivity – the spontaneous decay of certain unstable elements into different atomic species at a statistically predictable rate. Modern techniques of dating rocks using radioactive decay may be compared, albeit simplistically, to the use of the simple hourglass or egg-timer to measure the passage of time. The sand that collects in the lower bulb of the hourglass is, although imperfectly so, analogous to the daughter element accumulated by radioactive decay of the parent element, while the sand remaining in the upper bulb represents the amount of its radioactive parent element still remaining after a specific lapse of time. Highly precise methods of analysis, employing an instrument known as the mass spectrometer, yield the ratio of daughter to parent in the samples being tested, after which appropriate mathematical treatment will give the absolute age. The subject is a complex one that is not treated here beyond merely noting that it developed rapidly, and with increasing precision, after the pioneering work by Arthur Holmes in 1911, when he became the first man to date a rock by chemical techniques.

Holmes was the most far-sighted of all Earth scientists in Britain (and perhaps anywhere) through the first half of the twentieth century, and probably came closer than any other person who has ever lived to being 'the Compleat Geologist'. His *Principles of Physical Geology* (1965) is rated as the most successful geological book ever published. But *The Age of the Earth*, which appeared in 1927, was perhaps even more remarkable, considering the times in which it was written, and is certainly not to be judged by the price for which it sold – sixpence – in the days before inflation became a fact of life.

Holmes was greatly taken with Professor John Joly's theory that the history of our Earth reflects a series of 30-million-year cycles of radioactive heating to peaks culminating in volcanic or other igneous activity. Consequently, Holmes assigned ages to all the major periods back to the Cambrian as multiples of 30 million years. Checking these figures against the limited number of absolute ages available at that time, the Tertiary became 2 x 30 million years (m.y.) = 60 m.y., the Lower Cambrian 20 x 30 m.y. = 600 m.y., and the age of the Earth 1 600 m.y. to 3 000 m.y. The correspondence with our present-day numbers was remarkable – 60 m.y. as against 65 m.y. for the Tertiary Period, and 600 m.y. as against 544 m.y. for the Cambrian.

As the methods of rock dating became progressively more refined, the age given to the Earth continued to be stretched, pushing the 'beginning' further and further back into the distant past. By 1947,

J.H.F. Umbgrove at the Delft University was able to quote ten different methods, some based on radiometric analysis and others on astronomy. His best estimate was that 'the true age of the earth probably is between 2 and 3.5 billion years . . .' (Umbgrove, 1947: 13).

Using the isotopes of lead, ancient meteorites were in 1956 found by Claire Patterson in the USA to be close to 4.55 billion years old, and this was taken to be the age of the Earth. Until fairly recently, the rocks regarded as the oldest found on Earth were the Itsaq Gneisses of West Greenland (around 3 850 m.y.), and others occupying small areas in Labrador, Canada, China, Western Australia and Antarctica (Nutman et al., 2001). It had been mooted by Barton et al. (1983) that rocks exposed close to the border between the Northern (Limpopo) Province and Zimbabwe (the Sand River Gneisses) contain remnants with an age perhaps greater than 4 000 million (4×10^9) years, suggested at that time to indicate that 'they are among the oldest recognised terrestrial crustal rocks' (1983: 17).

The existence of even older rocks has now emerged from the discovery in Western Australia of zircons (crystals of zirconium silicate) caught up in sediments containing detritus shed by still more ancient source rocks. The following lines appear in the abstract of a paper by Wilde et al. (2001): 'Here we report, based on a detailed micro-analytical study of Jack Hills [Western Australia] zircons, the discovery of a zircon with an age as old as 4 404 ± 8 million years . . . This zircon thus represents the earliest evidence for continental crust and oceans on the Earth.'

There is a tale (possibly apocryphal) about a remark reputedly once made by a farmer while showing a visitor around his property in the Transvaal (now Limpopo Province): 'The rocks on my farm are 3 000 million and thirteen years old.' On being quizzed about where this curious figure came from, he replied: 'A geologist visited my farm and told me that the rocks here were 3 000 million years old. That was thirteen years ago!' This story reveals a popular misconception about both the accuracy and precision of rock-dating techniques – the better determinations of age (for example, Nutman et al., 2001) currently claim a precision in the order of 0.3 per cent.

A gap in the rock record, of at least 100 million to 200 million years, is perhaps to be expected, spanning the interval during which the earliest-formed crust of the Earth was in the process of forming, and probably subjected, from time to time, to re-melting. As Sir Archibald Geikie, Director of the British Geological Survey, put it in 1903, 'Geology by itself has not yet revealed, and is little likely ever to reveal, a portion of the first solid crust of our globe' (1903: 14).

An aspect of physics that leaves the mind somewhat bewildered is the conclusion that our Earth is in turn a relatively young body within the universe. In *Physics Today* of December 2003, the contributing editor P.F. Schewe summarised recent conclusions that gave the age of the universe as 13.5 ± 0.2 billion years. If there is one thing that emphasises the insignificance of humans within a geological context, this is it. Even when giving a generous allowance of 3 million years for humans' time span on Earth, we have featured in only the last 0.02 per cent of time elapsed since the birth of our universe.

The search for the age of our Earth has nicely revealed an axiom – consensus amongst the majority has at any given moment never been proof of anything. For several millennia the Earth had generally been taken to be less than 10 000 years old; for another 100 years consensus would have put the figure somewhere between 20 and 100 million years. During the first half of the twentieth century this figure crept into the billions, before it settled down to somewhere between 4 550 million and 4 600 million years. Is it at all conceivable that some unrecognised flaw in our concepts, such as that embedded in Lord Kelvin's calculations, will stretch this further at some time in the future?

3
Is the 'end' nigh? How much time do we have left?

'An old apocalyptic tradition fixed the duration of the world at six thousand years, analogous to the six days of Creation . . .'
Amos Funkenstein (1985: 56)

'. . . it was deducible that the duration of the earth as a planet must be strictly limited in time . . . when the store [of energy] is at length exhausted its life will cease, and it will become numbered amongst the dead planets.'
W.J. Sollas (1905: 301)

Wilbur G. Voliva, a minister in the town of Zion, Illinois, repeatedly predicted the end of the world. In his book, *A Hole in the Bottom of the Sea*, Willard Bascom (1961: 65) points out that the last time Voliva set a date for this event, he had predicted the year 1935, and adds the pithy comment that 'his forecast appears to have been wrong'.

Many gloomy predictions about the 'end of the world' have been made throughout history. While most dictionaries define the word 'apocalypse' (from the Greek *apokolypsis*) as an 'uncovering' or 'disclosure', in a wider sense the implication is one of impending catastrophe and doom. 'The apocalyptician,' writes Funkenstein (1985: 45), philosopher and historian, 'envisions the end of the world . . . convinces himself and his community that the end is inevitable and inevitably close'. Or, as Kermode (1985: 84) of Columbia University put it, 'the word apocalypse has no very precise meaning, only vague connotations of doom'.

UNFULFILLED PREDICTIONS IN LITERATURE

Visions of 'an end' would seem to have haunted humanity since the dawn of history. In his essay, 'A Schedule for the End of the World', Funkenstein's theme that 'every culture seems to harbour thanatic [Greek *thanatos* = death] fears of an ultimate catastrophe . . .' (1985: 45) takes us back to the apocalyptic visions in ancient Jewish and early Christian times. The calculation that the duration of the world would be a mere 6 000 years figures prominently in their literature. With a nice turn of phrase, Halpern (1998: 47) describes the story of Noah's Flood as 'apocalypse with an escape clause' – an event leading to something falling just short of total annihilation.

Written in the first century AD, is the awesome but savage imagery of 'an end' as envisioned in 'The Revelation of John' that brings the New Testament to a conclusion: in it we find the words '. . . the hour of fulfilment is near'. The Classicist Reiche, at the Massachusetts Institute of Technology, takes us even further back in 'The Archaic Heritage' (1985) to link, through the medium of archaeoastronomy, the preoccupation with catastrophe of the early Greeks and Romans, who connected it to eclipses of the Sun and Moon, and the slow drift (precession) of the constellations in the skies.

The Black Death – a plague spread by rat fleas – came as close to apocalypse as any other event in history. Beginning in central Asia, it raged across Europe between 1347 and 1350 AD, killing off an estimated total of more than 75 million people. Mercator, the cartographer responsible for one of the more commonly used map projections of today, predicted 1588 as the year of the apocalypse (Crane, 2003: 302).

Many people have put great faith in the *Centuries* of Michel de Nostradame (better known as Nostradamus), the mystic who has been described as 'everything from a Rosicrucian initiate intimate with the secrets of God to a drunkard whose badly rhymed quatrains are a dubious alternative to the proverbial pink elephant' (De Fontbrune, 1983: ix). Whatever we might think of the views of a sixteenth-century seer who is claimed to have predicted that the South African army would invade Russia at the end of the twentieth century (De Fontbrune, 1983: 417), Nostradamus was largely preoccupied with the demise of the Church, wars and 'worldwide conflagration which is to bring so many catastrophes and such revolutions that scarcely any lands will not be covered by water'. We cannot tell from his text whether he had in mind fire, another universal flood, or the collapse of organised religion, but his theme was that some massive catastrophe was scheduled for 1999 – another damp squib.

The year 1666 began in fear and trepidation in the Christian world, because the date included the mystical number 666, regarded by many as the symbol of the Antichrist (Brewer, undated: 790). The Great Fire of London came close to fulfilling the forebodings of the hundreds of thousands who were burned out of their homes in that year, lost all their possessions, and faced bankruptcy or starvation. Ninety churches, including St Paul's Cathedral, and 13 200 houses were reduced to ashes.

On the basis of his belief that the planets were fragments that had been torn from the Sun, the French physicist and mathematician, Comte de Buffon, had advanced the view in 1778 that the progressive cooling of the molten mass that was to become our Earth would end in total refrigeration under ice that covered its entire surface. He calculated the time for completion of this cycle from birth to death as 168 000 years (Geikie, 1905: 88), leaving some 93 000 years from the eighteenth century to the eventual extinction of all life.

According to a report in *The Weekly Telegraph* of 26 February 2003, Sir Isaac Newton was also obsessive in predicting the end of the world. Research by a Canadian academic, Stephen Snobelen, claims to have unearthed 4 500 pages devoted to the topic in manuscripts lodged in a library in Jerusalem. Newton reputedly predicted that the end of the world was due in 2060. We can only trust that this great physicist and mathematician will prove to be as wrong as have been all other harbingers of doom so far.

Figure 5: Visions of apocalypse, twentieth-century style, claiming that 'THE END IS AT HAND'. Yet another failed prophecy delivered to a disinterested crowd shopping in the famous flea market, Petticoat Lane, London, in 1965.

Friedländer (1985), a Professor of European history at Tel Aviv University, maintains that it was in the nineteenth century that, for the first time, the vision of the total destruction of humans by humans themselves appeared in the West. Against a background of pessimism in the aftermath of the French revolution and wars in Europe, strongly apocalyptic themes proliferated, and loomed large even in the works of leading popular writers such as Edgar Allen Poe, H.G. Wells and Jules Verne. The fundamental norms of civilised society (protection and championing of the weak or disabled) were questioned at that time: '. . . did they not threaten to interfere with [Darwin's] process of natural selection and provoke the progressive degeneration of man?' (Friedländer, 1985: 66).

Figure 6: Nothing is permanent in a geological context, even if it is of great age. Mukorob, or 'The Finger of God', was a freak of erosion that left a pillar of ancient sedimentary strata standing erect in the arid region of Namibia. The time elapsed between initial deposition of the sediments and the final sculpting of this remnant was in the region of 300 million years. There was an apocalyptic tradition amongst local communities that the world would come to an end if the pillar should fall – as it did in fact do before the end of the last millennium.

In a more modern context, the ultimate threat to humankind is seen by Harvey Brooks (1985: 112) of Harvard University to lie in thermonuclear war – 'a catastrophe besides which all other doomsday predictions are trivial'. His argument is difficult to disregard: 2 000 tons of nuclear missiles reaching non-military targets were adjudged to be capable of raising insuperable obstacles to the recovery of American society if hostilities should break out, at a time when an estimated 10 000 tons was held by each of the opposing superpowers. Other spectres raised by Brooks include the mutation of natural organisms to produce drug-resistant strains of bacteria or viruses, and a dramatic rise in the atmospheric concentration of carbon dioxide, a potent 'greenhouse gas' that would lead to massive global warming.

Another threat lies in the potential destruction of our protective ozone layer (Halpern, 1998: 37) resulting from our pollution of the atmosphere. The resulting flux of ultraviolet radiation would damage crops worldwide, leading to famine, and cause deformities in both animals and humans.

THE THREAT OF 'MISSILES' FROM SPACE

The above, small sample of predictions of doom reveals different concepts of what is meant by 'the end'. Some see it as the end of human life, while others perceive it as the end of all life. To some, the decline will be slow but progressive and inexorable, while others predict a decisive and catastrophic end. Models espousing biological mutations or global warming (notwithstanding forecasts of a mere 30 years ago that warned of our entry into a new ice age) are based on interpreting and extrapolating present trends. These contrast with cataclysmic models involving thermonuclear warfare, or geological/astronomical phenomena such as the collision of our Earth with a comet or meteorite to produce a world-encircling dust cloud matching or surpassing that which some assert (not entirely convincingly) wiped out the dinosaurs at the end of the Cretaceous Period (see more about this in Part 3).

Another possibility is that the entire planet could again be melted by impact with a colossal body from space, as postulated by Nisbet and Sleep (2001) to have occurred early in the history of our Earth. Some scientists speak not in terms of 'if', but 'when?' this might occur. A prediction emanating from the Harvard-Smithsonian Centre for Astrophysics in 1998, warning that a mile-wide asteroid, XF11, would have us looking at apocalypse in October 2028, was rescinded only after further calculations were made (Halpern, 1998: 35), but the possibility of collisions with 200-metre-diameter asteroids has popularised the phrase 'pushing the evolutionary Reset Button'. A group of scientists in the USA, known as the Spaceguard Survey Working Group appointed by the space agency NASA, has calculated that the chances of Earth actually being struck by a comet or asteroid larger than 1 kilometre in diameter are rated at about one collision every 300 000 years (Halpern, 1998: 33).

In 1908 the blast from an exploding meteorite destroyed over 5 000 square kilometres of forest at Tunguska, Russia. There are no grounds for us living in South Africa to assume that we are safe from violent phenomena such as this. Of the approximately 170 impact craters known on the face of the Earth in 2001, at least eight significant examples have been recognised in southern Africa. The greatest of them all is the Vredefort

Ring Structure, south-west of Johannesburg and traversed by the Vaal River, where the ancient sequences of the gold-bearing Witwatersrand succession have been thrown into a ring-like form, some 80 kilometres in diameter, in which the strata now stand vertically, or are locally even heaved upside down. The core of the structure, which punched its way upwards into the Witwatersrand and overlying strata, is occupied by ancient rocks of the Basement, believed to represent exhumed original crust that had previously lain at a depth of between 8 and 11 kilometres below the Earth's surface.

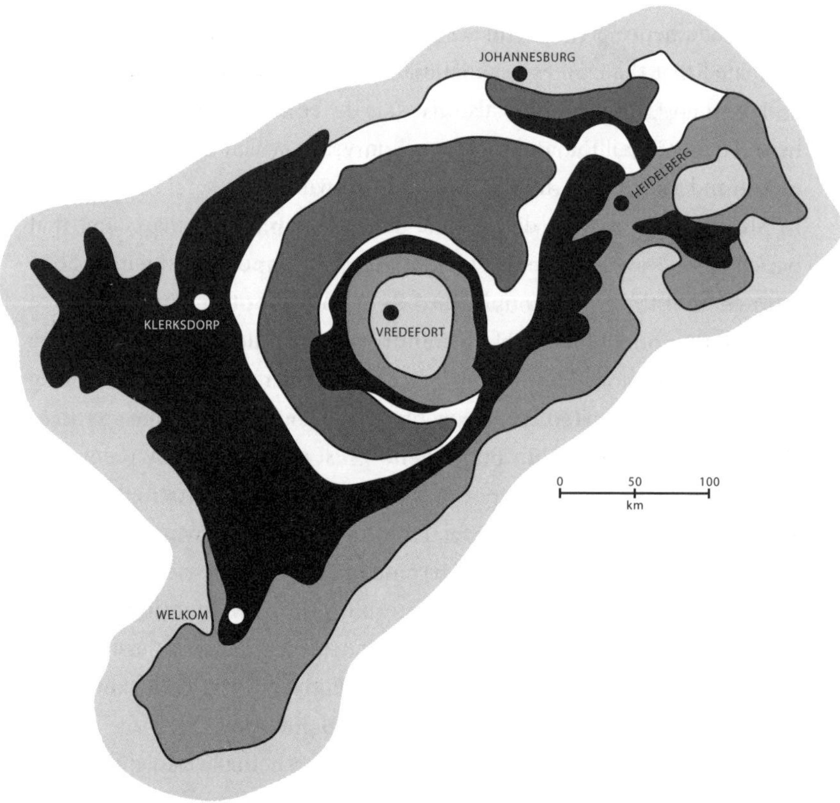

Figure 7: A simplified geological map (adapted from that published by Gibson and Reimold, 2001) showing the great basin of Witwatersrand Supergroup rocks (medium grey) resting upon ancient Basement rocks (pale grey), and overlain by the Ventersdorp (black), and lower (unshaded) and upper (dark grey) Groups of the Transvaal Supergroup. The essentially circular form of the Vredefort Dome is clearly defined in the centre of the region, where the central core of Basement granite has punched its way upwards to tilt the surrounding strata into near-vertical attitudes. Much younger Karoo Supergroup rocks that cover the southern part of the region have been omitted to simplify the diagram.

The origin of this extraordinary structure has been a highly controversial topic for close on 100 years. Some earlier explanations attributed it to conventional 'endogenic' processes (stresses generated and acting within and below the Earth's crust). Others linked it to the intrusion of magma emanating at depth from the Earth's mantle. General consensus is, however, now shifting towards accepting new lines of mineralogical, petrological (pertaining to the origin, composition and texture of rocks) and structural evidence, summarised by Gibson and Reimold (2001), that indicate that the feature was created by a huge meteorite impacting on Earth at a velocity in the order of 25 kilometres, or more, per second. Although now deeply eroded and hence greatly reduced in size since the moment of impact, it is estimated that the crater's original diameter was 250 to 300 kilometres. A variety of applied techniques also gives us the age of the event, somewhere between 2 002 million and 2 023 million years, which makes it both the oldest, and the greatest, impact structure known on Earth.

It would be easy to dismiss the Vredefort event as an accident that occurred so far into the past that it merits no further consideration. The bad news is, however, that it is only one of a train of events irregularly spaced through time, and that one of the more recent significant impacts in South Africa took place a mere 220 000 years ago. The resulting structure, once known as the Pretoria Saltpan (now the Tswaing Crater – see Chapter 15 of this book), lies 40 kilometres north-northwest of Pretoria (now Tshwane). It has been calculated that the meteorite that struck here would have been some 30 to 50 metres in diameter, had impacted at a velocity of 20 to 30 kilometres per second, and would have released the same amount of energy as 100 Hiroshima-type bombs. If the release of this amount of destructive energy was capable of producing a crater with a present-day rim-to-rim diameter of only 1.13 kilometres, how much greater would have been the output at the Vredefort site, with a diameter of 300 kilometres?

Another comparison might be drawn with the Chicxulub crater on the coast of the Yucatan Peninsula of Mexico. The diameter of the transient crater produced at the moment of impact there has been estimated to have been 90 to 105 kilometres before collapse of the strata surrounding it widened it to 180 to 210 kilometres (Snyder & Hobbs, 1999). Although this is appreciably smaller than the Vredefort structure, many scientists hold the view that this was the event that triggered the extinction of numerous life forms (including the dinosaurs) at the end of the Cretaceous Period, 65 million years ago.

WHAT IF WE SHOULD LOSE OUR PROTECTIVE MAGNETIC SHIELD?

In recent years increasing coverage has been given to another phenomenon – the decay of the Earth's magnetic field. If the majority of the world's people have given any thought to this at all, it is probably in terms of some mysterious force that provides no more than a method of navigation on the surface of the Earth, using a magnetic compass. There is, however, a rather more vital function – the Earth's magnetic field deflects dangerous radiation emanating from our Sun. If the Earth were to be deprived of this protective umbrella, all life would be under threat. There is no plausible way of quantifying this danger, but phrases such as 'deaths in the hundreds of millions of people' are used by some scientists.

The Earth's magnetic field originates in the liquid part of its core (see Chapter 4), and there is a quirk in that this magnetic field periodically reverses itself, that is, the North and South magnetic poles change places at irregular intervals. This does not imply that this is an event that might occur overnight, or that in the days before geographic positioning by satellites, ships destined for Europe would have been sent crashing into Antarctica. Hundreds to thousands of years are required for the change to be completed.

The physics underlying these changes is not relevant here, but the history of the reversals that is written in the rocks of our Earth's crust most certainly is. When igneous rocks cool from their parent molten magmas, they preserve (as they pass through what is known as the Curie temperature) a record of the Earth's magnetic field prevailing at that moment. Magnetic particles in sedimentary rocks retain similar information by orienting themselves in the magnetic field as the sediments accumulate. From the compilation of available geological data, it is known (for example, Glatzmaier et al., 1999) that there have been several hundred reversals of the Earth's field during the last 160 million years alone. In addition, there have been 'geomagnetic excursions', which are deviations of more than 45 degrees of latitude from the geographic poles without achieving full reversal – perhaps fourteen of them in the last 780 000 years.

What is important is that the record shows that reversals are preceded by a progressive decay in the strength of the magnetic field – an ominous finding in view of what Olsen (2002: 594) describes as 'the precipitous decline of the Earth's main dipole field over the past 150 years'. The implication is that a reversal may be under way at the present time and, at the current rate of decline, the magnetic field may well vanish early in the next millennium.

When comparisons with planet Mars are made, it is perhaps time to sit up and take notice. Mars has no magnetic field, but residual fields preserved in its rocks show that at one time such a field did exist. As it is smaller than Earth, Mars may have cooled more rapidly and allowed its liquid core to solidify, thus losing the capacity to generate a magnetic field. The absence of a Martian atmosphere might then perhaps be attributed to its having been blown away by the solar wind, leaving behind an extremely hostile and effectively 'dead' environment. Certainly, no firm predictions can be made about Earth's magnetic field or the effect that its (even temporary) loss would have, but it is yet another scenario that falls within the category of apocalyptic forecasts.

If we take a wider view, however, the hazards attributable to magnetic pole reversal may perhaps be perceived as exaggerated – by and large, life seems to have survived the several hundreds of earlier events written into the rock record. Perhaps the loss of our ozone layer, so much in the news at present, and the surge of ultraviolet radiation that would follow, is a matter of greater urgency. It is also something that falls within our power to ameliorate to some degree, unlike the magnetic field, over which we have no control whatsoever.

We are left, therefore, with a riddle. Is the person who walks the streets displaying a sign such as 'Take Heed – the End is Nigh' to be taken seriously or not? All that we can say is that physics, geology and astronomy suggest that he or she is probably right, even if for the wrong reasons, but the time when 'the end' will come is beyond our competence to predict.

4
Pluto's realm: What lies at the Earth's centre?

'. . . those who go down to the deep in mines know well that the temperature rises as we descend . . . if we consider that every cubic foot of earth weighs perhaps one hundred and fifty pounds or more, it is certain that the heat some ten or twenty miles below the surface must be very great, and must be still greater further down, but yet every evidence goes to prove that . . . the interior of the earth is kept solid by the enormous pressure.'

Sir Charles Metcalfe (1904: 15), presidential address to the South African Association for the Advancement of Science

It is no simple matter to assess what lies deep within our Earth. Direct observation in our deepest mines is limited to about 4 kilometres, while some boreholes have been driven down more than 12 kilometres. But, with the Earth's radius being close to 6 400 kilometres, these holes are no more than pinpricks. Constructing a picture of the deep interior of the planet can therefore be based only on indirect evidence – and here there has been, over the years, scope for the most startling differences in interpretation.

BEGINNINGS IN MYTHOLOGY

It is quite extraordinary that the ancient Greeks, without any knowledge of geophysics or geology in the modern sense, had equated increasing depth with fire and great heat. The realm of Hades was their 'nether world' into which the spirits of the dead descended. In later Roman mythology, this realm was presided over by the god Pluto (hence the term 'plutonic rocks' in modern geology, signifying their generation at depth and at high temperatures). We can surmise that it was the contrast between the high mountains covered with snow, and the hot ash and molten lava spewed out from the throats of volcanoes, that led early people to the notion that 'up is colder, and down is hotter'. The traumatic burial of Pompeii and Herculaneum under volcanic ash in 79 AD would likely have confirmed this belief.

According to Healy (1986: 112), it was the ejection of molten rock from the volcano of Mt. Etna that prompted Empedocles (492–432 BC) to argue that the Earth's interior must be in a molten state. Through the ages, the merging of scriptural visions of Hell with those of the 'nether world' has preserved this notion of raging infernos somewhere at depth. The most extreme view of this could be that of the mystical Jewish Cabbalists who, according to Brewer (undated: 540) imagined seven successively deeper levels, the first being at 60 times the temperature of fire, and the last at 60 to the power of 7 (60^7), which equates to 2 799 billion times the heat of fire.

John Milton the poet, too, made his contribution in *Paradise Lost* (1674) in his vision of the '. . . Dungeon horrible, on all sides round/as one great Furnace flam'd . . .' (Bk I: lines 61–62) without being exactly explicit about where this was located. Nevertheless, his mindset is implicit in the phrases: 'Ætna, whose combustible/And Fewel'd entrails thus conceiving fire . . .' (Bk I: lines 232–34), 'this deep world of darkness' (Bk II: lines 262–63), and '. . . long is the way/and hard, that out of Hell leads up to Light' (Bk II: lines 432–33).

THE EVOLUTION OF IDEAS: FIFTEENTH TO NINETEENTH CENTURIES

Apart from some notably bizarre ideas, the belief in the Earth as a body whose temperature increases with depth has survived through the ages. In Agricola's *De Re Metallica* (1556), we read his explanation for the deposition of metallic ores by infiltration of 'subterranean waters' emerging from the depths possessing 'subterranean heat'. We recall, also, Leibnitz's notion in 1680 of the Earth cooling from a molten mass, and Comte de Buffon's calculation in 1776 of the time that it would have taken a molten body of this size to cool to its present condition.

James Hutton (1726–97), the Scottish doctor and farmer whom many see as the true 'father of geology', is credited by Repcheck (2004) as the man who took the first step on the path that led ultimately to the modern theory of Plate Tectonics (see Chapter 6). While this may be a somewhat fulsome attribution, no one could quibble with Repcheck's claim that 'though other scientists had recognised that the inside of the earth was hot, it was Hutton who saw this as crucial to the earth's overall workings' (2004: 211). To Hutton, it was heat in the Earth's interior that was the ultimate cause of the uplifting of continents and the floors of the oceans, and of the recrystallisation of rocks during metamorphism.

As late as 1897 Lord Kelvin's belief in the Earth's internal heat being a leftover from a time when it was a molten mass was quite acceptable. That was the rationale for his calculation of the Earth's age. Then came the discovery of radioactivity, which revealed that this model of a steadily cooling planet was totally unacceptable. The cumulative effects of radiogenic heating were recognised as being capable of raising rock temperatures to melting point throughout the Earth's history, because of the poor conductivity of rocks – their inability to conduct heat efficiently and radiate it into space.

In the *Text-book of Geology*, one of the definitive books of the time, Sir Archibald Geikie (1903: 56–74) concluded that 'no fact is of more importance than the existence of a high temperature beneath the crust, which has now been placed beyond all doubt'. He backed this up by pointing to volcanic eruptions, hot springs yielding water close to boiling point, and elevated temperatures in deep mines. From a plethora of models in vogue at that time, Geikie selected only three as worth considering: solid crust overlying a molten or viscous interior;

solid crust overlying solid material to the centre of the Earth; and solid crust overlying liquid magma enclosing in turn a gas core under enormous pressure and high temperature. The first of Geikie's three alternatives came closest to our modern perceptions of a rigid crust, 25 kilometres thick on average, overlying a viscous mantle subject to periodic melting, and a metallic core at depth, liquid in its outer part, but solid in the centre.

SCHWARZ'S COLD EARTH

Even allowing for the immaturity of ideas a mere 100 years ago, many of the hypotheses offered at that time were somewhat bizarre. Ernest Schwarz, a graduate of the Royal College of Science and the Camborne School of Mines before joining the Cape Geological Commission, adopted, metaphorically, the role of a player with a trumpet tuned to a key that was different from the rest of the orchestra playing Gustav Holst's *Planet Suite*. Boldly Schwarz asserted (1909) that 'all the so-called proofs and reasons for the earth's interior being hot and liquid rest on arguments which will not hold water', before coming out with his conclusion that:

> . . . the centre of the globe is cold . . . the slice of the earth represented by South Africa is being warmed by radium by an amount equal to the consumption of 9 million tons of coal a day. This rate of evolution of heat is sufficient to explain the whole of the temperature gradient in the earth's crust. . . . If the centre were hot the added heat would render the earth's surface uninhabitable (1912: 17–19).

Schwarz's conviction would seem to have begun with a misinterpretation of geophysical data available at that time. Seismic (earthquake-generated) energy is transmitted through rocks by waves of several different types. He was clearly familiar with the principle that the so-called 'P-type' compressional waves (like sound waves) are transmitted through liquids whereas those of 'S-type' (shear or transverse waves) are not. His observation that waves of both types penetrated through the rocks underlying the Earth's outermost crust had led him, therefore, to reject the concept of the Earth having a hot, liquid core. He had assumed that the physical properties of rocks immediately beneath the crust would be much the same as those of the deeper interior of the Earth as well. He therefore seized upon T.C. Chamberlin's 'Planetismal Hypothesis' of the Earth growing by the accretion of solid matter floating in space at a temperature only slightly above absolute zero (–273 °C). The obviously higher temperatures of the rocks of the crust, on which we now live, he attributed to heat emitted by the element radium.

In all fairness to Schwarz, we should allow that he did temper his eccentric hypothesis with a defensive remark that appeared in 1912 in the Preface to the text of his book, *South African Geology*: 'I have, however, put in some material which is still on the side of unrighteousness; for example, the Planetismal Hypothesis.'

Judging from his numerous publications, and the reminiscences of those who knew him, Schwarz was not a man who easily changed his mind. He reasserted (1912: 19) his conclusion that Earth temperatures would peak at a depth of 'not more than 20 000 or 25 000 ft. [6 100 or 7 600 m] below the surface', and decline from there to a centre that was as cold as outer space. Calculation reveals that the peak as he saw it would be reached at a mere 260 °C. How then, was this to be reconciled with the occurrence of molten magmas of igneous rocks, including incandescent lavas discharged by active volcanoes?

This paradox led Schwarz into a discussion (1912: 57–59) suspiciously

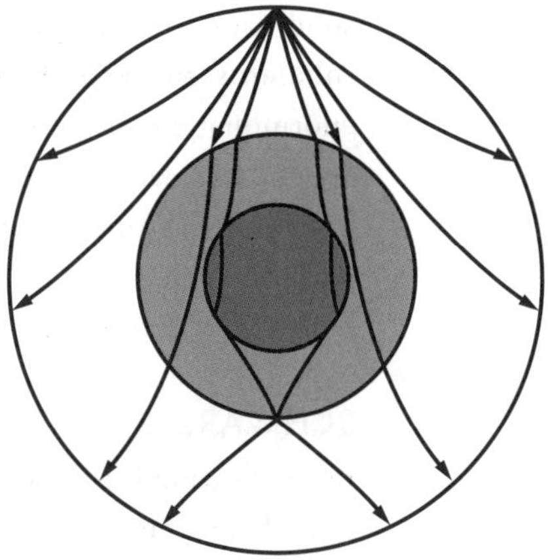

Figure 8: Cutaway view through the Earth showing the thin outer crust (outermost circle), mantle (unshaded), outer liquid core (light grey), and inner solid core (dark grey). The curved paths followed by earthquake P-type or compressional waves, with refractions at major boundaries, show how the internal structure of the Earth was deduced. S- or shear-type waves do not penetrate the outer core, from which it is reasoned that this must be in a molten state, as S-waves cannot be transmitted by liquids. The thickness of the crust ranges from a mere 5 kilometres to around 70 kilometres; the base of the mantle is at approximately 2 900 kilometres; and the Earth's radius at the Equator is 6 378 kilometres.

Figure 9: Residual heat given off by relatively young igneous rocks emplaced near the surface by magmas rising from the Earth's mantle may heat the Earth's crust. In the 'Craters of the Moon' geothermal area near Lake Taupo, North Island, New Zealand, scalding steam issues from the soil, and the area is dotted with pits of churning, boiling mud.

redolent of the archaic ideas of Abraham Gottlob Werner (1749–1817) of Freiberg in Saxony. Werner was the founder of the Neptunist theory that the oldest rocks of the Earth's crust had been chemically precipitated from a primeval ocean (see Chapter 1). In an abstruse presentation, Schwarz (1912: 57) submitted that 'there is a large body of evidence which goes to show that [the parent liquids of igneous rocks] are rather solutions of substances in one another'. The temperatures of such 'solutions' were not discussed, but Schwarz asserted that 'though heat quickens the process it is not essential'. Where the molten lavas of volcanoes provided irrefutable proof of high temperature, he shrugged this off by claiming that this was generated by local frictional heat resulting from movements within the outermost 10 to 12 miles (16 to 20 km) of the Earth's crust.

A HOT CORE – BUT SOLID, LIQUID OR GASEOUS?

Fourteen years after Sir Charles Metcalfe's 1904 presidential address to the South African Association for the Advancement of Science (quoted at the start of this chapter), Professor J.T. Morrison had weighed in with conclusions very similar to those of his predecessor in office. His scenario combined Lord Kelvin's basic premise that Earth temperatures increased to around 4 000 °C in the core of the Earth, with his own view that 'S-type' seismic waves did indeed pass through it. His text leaves us in no doubt as to his conviction that:

> the deduction made by most geologists and many physicists till well on in the nineteenth century was that a part of the earth must be liquid, and the innermost part a highly compressed gas . . . There are, however, many facts which militate very strongly against the soundness of such conclusions (1918: 35)
> . . . It is impossible to regard any continuous shell of it as being liquid (1918: 39).

The beginning of the twentieth century was thus a time of uncertainty about the nature of the Earth's interior. Was there a hot core, cold core, solid core, liquid core or gaseous core? Could the Earth be compared to the old-fashioned golf ball, with a tough outer skin enclosing a thick layer of viscous material wrapped around a central sac of some dense liquid?

LATER MODELS

Since the early years of the twentieth century, much has been learned about the structure of the Earth, and current models are very different from that conceived by Schwarz. Modern views are based on a wide spectrum of observations, such as the interpretation of earthquake waves, the flow of heat through rocks, and the levels within them of the heat-producing radioactive elements uranium, thorium and potassium. Measurements of the melting temperatures of common igneous rocks have revealed a complex interplay between temperature, pressure and the presence of fluids (for example, water) that leads to their melting, but point to a range, in very broad terms, of temperatures in the region of 700 °C to more than 1 300 °C being required to keep different types of magmas liquid.

The chemical composition of common basaltic lavas is now known to fit the generation of their magmas by the melting of rocks of the mantle at depths of 40 to 70 kilometres. Others (for example, kimberlite magmas, host of the precious diamond) are generated at much greater depths in excess of 150 kilometres, while the dying out of earthquake foci at depths around 600 kilometres points to a softening of originally shallow, brittle rocks thrust deep into the mantle by the forces that drive continental drift.

Average geothermal gradients (the rise in temperature with depth) are now taken to be close to 10 °C for each kilometre of depth down to 100 kilometres beneath continents, and 13 °C per kilometre beneath oceanic areas (Wyllie, 1971: 30). Gradients are steepest near the Earth's surface, but decline as depths increase (see Figure 43), with temperatures in the core of the Earth estimated to be in the region of 5 000 °C. The interpretation of seismic data and the physics of the Earth's magnetic field indicate that the outer part of the core of the Earth is largely liquid (molten iron plus nickel), but that this encloses a further inner sphere of metal solidified by the enormous pressures.

Without the gradual evolution of this basic model of the Earth, much of what we have learned about our planet in the last 100 years would not make much sense. We would have no explanation for the way in which continents have in the past drifted on the surface of the Earth (and still do at a rate of several centimetres each year). Without the high temperatures of the rocks beneath our crust, the slow convection of currents that carry the continents on their backs would be impossible. We would have no reply to those who ask how volcanoes come into being, or how the great lava fields of the world were spilled out on its surface, or how injections of molten magma into the crust were possible, to form intrusions such as the huge Bushveld Complex (see Chapter 14) in South Africa. Without a liquid core, there would be no magnetic field to give us protection from solar radiation, with the result that our Earth would be as 'dead' as Mars.

TO EVERY ACTION A REACTION

All but a small number of scientists would these days dismiss Schwarz's 'cold Earth' as a highly eccentric idea. As a tailpiece, however, it is worth noting that objections to the Plate Tectonic Model voiced at a 'New Concepts in Global Tectonics' meeting held in Urbino, Italy (Wezel, 2005), as part of the thirty-second International Geological Congress (and in celebration of the five-hundredth anniversary of the University of Urbino), revived the concept of an expanding planet within which both the Earth's mantle and core are cold and solid. Wezel concluded that the theories presented offered 'scientific concepts which highlight various aspects of a many-sided reality of great geological complexity'. Could this be the rebirth of an old controversy, like Phoenix rising from the ashes?

5
Is the Earth expanding or shrinking?

'The favourite conception was one that invoked the contraction of the earth in consequence of its cooling. Thus Cordier calculated that a contraction of only a single millimetre would suffice to force out to the surface lava enough for 500 eruptions, allowing 1 cubic kilometre for each eruption.'
'Mallet contended that variations [in the geothermal gradient] are too great . . . to be due merely to differences in the transmission of internal heat . . . they point to local accessions of heat . . . due to the constant cooling and contraction of the globe.'
Sir Archibald Geikie (1903: 351, 64)

'There is now ample evidence for seriously considering the hypothesis that expansion of the interior of our planet may have played a dominant role in geological history and the evolution of surface features . . .
Present-day stress fields, inferred from seismic records and also directly measured, favour expansion rather than contraction.'
Arthur Holmes (1965: 965)

'If it was not certain whether the earth was heating or cooling, contracting or expanding, whether continents were drifting or not, perhaps the only common-sense approach was to be mute.'
H.W. Menard (1986: 82) describing the lack of consensus in 1930–40

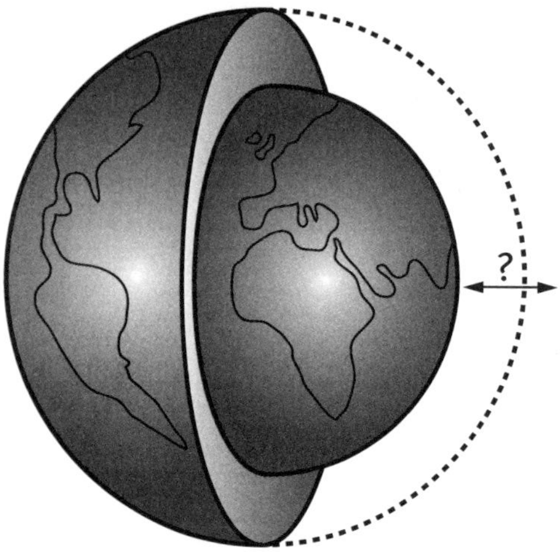

Figure 10: The Earth: expanding or shrinking?

It would be difficult to find two points of view more opposed than those enunciated by Geikie and Holmes. There could be no compromise, with one faction claiming that the Earth is shrinking and the other that it is expanding. The aim of this chapter is, however, as much to present the opposing views about this as to illustrate how the concepts in the nineteenth and early twentieth centuries became woven into a variety of different explanations for other natural geological phenomena.

A SHRINKING EARTH?

The conclusion that the Earth was at one time a molten sphere that cooled to allow the growth of a solid outer crust, was the foundation from which arose the belief that the radius of the Earth must be shrinking. The way in which rocks are intensely folded and contorted in the great mountain chains of the Himalayas, the Alps and elsewhere seemed to require a general shortening of the circumference of the Earth – in much the same way as the skin of an apple will wrinkle if it is left to dry out. This basic premise led to a philosophy that was, 100 years ago, profoundly different to that of today.

As one example, there was the proposal of Robert Mallet in 1873 recalled by Geikie (1903: 352). Mallet was one of the founders of seismology, and probably one of the first people to use gunpowder to generate shock waves artificially in order to measure the speed at which they passed through rocks. His conclusion was that volcanic activity at the surface of the Earth was the consequence of 'the more rapid contraction of the hotter internal mass of the earth and the consequent crushing of the outer cooler shell or crust'. This crushing he held to have taken place along lines of weakness, and 'it is precisely along such lines that crumpled mountain ridges and volcanic funnels should appear'. Mallet argued that the heat produced by this crushing would be capable of reaching temperatures at which actual fusion of rocks would take place. Then, if water should find its way down into such zones, it would flash into steam and result in what he described, in the florid language of the times, as the 'commotion of a volcanic eruption'.

Another result of this doctrine (Geikie, 1903: 370) was that 'there is now general agreement that the cause of important earthquakes is to be sought in the effects of terrestrial contraction'. It was concluded that any part of the Earth's crust under compression might suddenly give way, and 'under the intense strain . . . snap along the length of the anticlines or synclines [folds], and thus give rise to the tremors or more violent shocks of mountainous regions . . .'

AN EXPANDING EARTH?

Controversy was inevitable, for others interpreted matters very differently. In 1933, O.C. Hilgenberg, a supporter of the idea of continental drift, had demonstrated that the globe could be reconstructed with the continents fitting together to cover its entire surface in the past – if it were to be accepted that the Earth's radius had increased from 4 000 kilometres to its present value of 6 378 kilometres (an increase of some 50 per cent). The President of the South African Astronomical Society, J.K. Halm, had estimated that an increase of 1 000 kilometres in radius would account for the splitting apart of the continents and formation of the ocean basins. Even more extreme was the much-publicised estimate of S. Warren Carey of the University of Tasmania. In 1958 Carey theorised that the radius of the Earth had doubled, and that the rate of increase was actually accelerating.

In his best-selling book, *Principles of Physical Geology*, Arthur Holmes (1965: 965–83) gathered together various threads that led him to conclude that, far from contracting, the Earth is actually growing larger. This he attributed to the core of the Earth undergoing a change of state (from what physicists termed dense, pressure-ionised matter) to material of greater volume. By his reckoning, estimates of the amount of 'crustal shortening', one of the main props of the dogma of contraction, were exaggerated.

Holmes revived the ideas of L. Egyed, Director of the Geophysical Institute at Budapest, who had looked at the problem from a different angle. Egyed had estimated the total areas of continents that had been covered by the sea at various times since the Cambrian Period (around 540 million years ago) and found in these estimates a progressive decline with time. This was what would be expected of an Earth that was increasing in size, while the volume of water available in the oceans remained the same. Calculations based on these reconstructions had suggested to him an increase of the Earth's radius by a little less than half a millimetre per year.

Holmes also looked at the problem from another angle – the gradual lengthening of the day. From astronomical observations, this is about 2 seconds per 100 000 years, as the Earth's rotation on its axis slows down. A basic principle in physics – the conservation of angular momentum – requires that if the Earth should expand without any change in its total mass, the speed of its rotation must decrease.*

Confirmation of this slowing down (although not of the cause) came from a surprising source – palaeontology. Growth rings in the skeletons secreted by Devonian and Carboniferous fossil corals had been identified by J.W. Wells of Cornell University as daily growth-lines, and counting these gave an estimate of the number of days in the year at those times: 390 to 410 shorter days. From physics, this could again be interpreted to mean an increase of the Earth's radius by approximately 0.6 millimetres per year.

The model of an expanding Earth threw up new problems of its own. Expansion would imply a crust under tension (that is, stretching, as with a balloon being inflated), whereas the crumpling of strata that is characteristic of mountain building would seem to require compression (a 'squeezing together'). To meet this apparent contradiction, Egyed had argued that as the Earth's radius had increased, the continents would have had to deform and wrinkle in order to adapt to a lesser degree of curvature on the surface of a sphere. This deformation would have caused localised compression. In review, one of the American pioneers in developing the Plate Tectonic Model, H.W. Menard (1986: 145), dismissed the entire concept as 'a quicksand of misinformation'.

The review by Menard (1986: 65) shows that as late as 1950 there was no consensus on contraction of the Earth as opposed to expansion: 'He [Beno Gutenberg] went on to show that the hypotheses of crustal contraction, subcrustal convection, polar wandering and continental drift of some sort were still viable at that time.' Evidence that could not be refuted seemed to support each of the models, and Menard remarked that 'global tectonics was in a sorry state'. Even J. Tuzo

* An analogy often used by teachers of physics is that of an ice skater spinning on one spot during a skating routine. With the arms at first held to the sides, the speed of rotation decreases rapidly as the arms are raised to shoulder-height and extended sideways. There is, however, also an alternative explanation for the slowing down of the Earth's rotation – gravitational interaction with the Moon.

Wilson, the American who would within a few years become an ardent supporter of the concept of the spreading of the sea floor, had been a contractionist who in 1959 estimated that a shrinkage of 150 kilometres in the circumference of the Earth was a viable hypothesis.

There were profound implications to be considered in the concept of an expanding Earth. A doubling of its radius would have resulted in an eightfold decrease of its average density, meaning that this would have been in the region of 44 grams per cubic centimetre in the past (to be compared with the present value of 5.54 gms/cm^3), and the pull of gravity on Earth's surface therefore four times that of the present time. Opinions differ on the effects that a gravitational field as powerful as this would have had on the evolution of life forms or, indeed, on most biological and geological processes – the circulation of blood in animals, or water in plants; the development of muscle and bone of animals; rates of erosion on the surface of the Earth; and rates of sedimentation in oceans, lakes or rivers.

So much for the concept of Uniformitarianism!

THE CONTROVERSY RESURGENT

Perhaps old arguments never die, or, as one mischievous humorist has put it, 'old academics never die, they merely lose their Faculties'. The Plate Tectonic Model (see Chapter 6) had seemed to be so complete in its ability to explain the features of our Earth (continental drift, earthquakes, volcanic phenomena, etc.) that the controversy of contraction versus expansion had faded into the background for several decades. However, it has recently been stirred to life once more.

After the 'New Concepts in Global Tectonics' meeting already referred to, held in Italy in 2004, Wezel (2005) reported that participants from many different countries had met with the expressed purpose of '. . . overcoming the mental bias and resistance linked to the assumption of the monolithic paradigm [that is, the Plate Tectonic Model] which has dominated the global tectonic scene in the last 37 years . . .' In one contributed scientific paper after another, the case for an expanding Earth was revived and developed. The fields covered included seismic (earthquake) studies, volcanism, the origin of mountains, and plumes of heated rock rising from the Earth's mantle. The claim was made that the core of the Earth is cold, but is continuously adding to the volume of the mantle, atom-by-atom, thereby increasing the radius of our planet. Only time will tell how much support these arguments garner.

6
CONTINENTAL DRIFT: THE GREATEST CONTROVERSY OF ALL

'The many similar forms of life, either fossil or recent, that are found scattered over various parts of different countries now so widely separated by the Indian and Pacific Oceans, seem to indicate that in very remote periods they must have been more intimately connected with each other than they are at present.'

G.W. Stow (1870), self-taught geologist writing about the Eastern Province of South Africa

'It may seem a bold undertaking to connect Africa and Brazil by a continental mass of land in order to explain a simple fact in animal distribution; but the independent investigations of Suess and a great number of other thinkers have afforded much evidence in support of the supposed extension of the Mediterranean into the Caribbean sea at a period when neither North nor South Atlantic, at least in their present conformation, were in existence.'

W.J. Sollas (1905: 283)

'There is . . . not the slightest reason to believe that bodily displacements of continents through the lithosphere are possible . . . Secular drift of continents . . . such as have been maintained by A. Wegener and others, are out of the question. A small drift of the crust as a whole over the interior is not impossible . . . but the maximum amount seems to be too small to be of much interest.'

Sir Harold Jeffreys (1929: 305), Cambridge mathematician

'Since such a principle receives the fullest support from every other line of evidence . . . it is insisted in conclusion that the Hypothesis of Continental Drift must now be regarded no longer as a mere speculation, but as a definitely established and fundamental truth . . .'

Alex L. du Toit (1937: 331)

'In certain quarters the Theory of Continental Drift had come to be regarded as being in the nature of a monstrosity inducing abhorrence and nausea, even of a moral turpitude unmentionable in "nice" geological society. Adherence to the new heresy immediately put one outside the pale of rational human beings, to say nothing of refined intellect. Any attempt to show "that there might be something in it", was treated with scorn and derision and led to reflections on the adequacy of one's decent bourgeois upbringing . . . The theory of continental drift probably has passed its zenith. It may eventually be entirely rejected.'

T.W. Gevers (1949: 84–85, 102), Professor of geology, University of the Witwatersrand

'One perceives ever more clearly the total vacuousness and sterility of the hypothesis . . .'

V.V. Beloussov (1962), Academy of Sciences of the USSR, as quoted by Menard (1986: 186)

'The evidence of palaeomagnetism, with which that of palaeoclimates does not conflict, suggests that the poles of rotation of the earth and the land masses have gradually changed their relative positions . . . The existence of this relative motion in the earth's crust today implies that movements deeper in the crust are taking place for which we have no adequate theory.'

S.K. Runcorn (1959) as quoted by White (1962: 287)

'There is now general agreement that palaeomagnetism was the most convincing evidence set forth in support of the concepts of continental drift and sea-floor spreading . . . The implications of plate tectonics are so far-reaching that this theory can be considered the framework from which to view most other geologic processes.'

P.J. Tarbuck and F.K. Lutgens (1984: 398, 400)

Ever since Charles Darwin's startling ideas on evolution made the headlines in the 1850s, and were (mis)interpreted by the world at large to mean that humans were descended from a monkey, scientists have been looked at somewhat askance. Did the Bible not describe how the first humans were created? To make matters worse, the formidable Thomas Huxley had given spirited support to Darwin's ideas on evolution, and had been widely quoted in the press.

This rather sceptical view of science was hardly improved in the early twentieth century by a seemingly ridiculous idea that, in the distant past, the southern continents of the world had been united as one enormous supercontinent. Was it to be believed that South America, the Falkland Islands, Africa, Madagascar, Antarctica, peninsula India, Australia and New Zealand had once fitted together like a gigantic jigsaw puzzle, simply because of some coincidence of their shapes? Ostensibly all of this was based on the distribution of the leaves of an obscure fossil plant *Glossopteris* (see Plate 2), of reptiles, and of now widely scattered rock sequences showing the same transition from glacial to temperate, and then to hot, desert-like climatic conditions. Had this warming up created one of the greatest deserts in recorded Earth history, spreading across South America then united with Africa, and was all of this brought about by the northwards drifting of the supercontinent away from the South Pole? Even to give to this imaginary (if not fictitious?) landmass the outlandish name of Gondwanaland had seemed presumptuous.

The idea would have seemed preposterous to those who were being asked to believe that the golden beaches of KwaZulu-Natal, or the rugged cliffs of the Transkei, have not always stared out over the blue Indian Ocean. In this scenario, there would have been no South Atlantic Ocean to lap the bays and islands of Rio de Janeiro several hundred million years ago, much less to suggest the future location of the famous playground of Copacabana Beach, or of a place where there would one day be a song written about the 'Girl from Ipanema' in Brazil. Furthermore, whatever time it might have taken to travel by coach, on horseback, or on foot, between Sydney and Perth, it would have taken close on five times longer to travel from eastern Australia to the west coast of South America across this enormous land mass, *without crossing any great ocean*. To a sceptical world, this creation from the minds of some scientists would have appeared about as real as the legend of Camelot and King Arthur.

THE GROWTH OF AN IDEA – AND THOSE WHO SCORNED IT

Scientists cherish the aura of reaching their conclusions through dispassionate logic and hard facts. Some even believe in this myth. However, the liberal sprinkling of phrases such as 'a monstrosity inducing abhorrence', 'new heresy', and 'total vacuousness and sterility of the hypothesis' in the quotes at the start of this chapter is more suggestive of faces flushed with indignation and wounded pride than of calm debate. Indeed, that is how it was, for of all the great controversies that were the weft and warp of threads in the fabric of modern geology, continental drift was the greatest. In the weaving of it, there were at first so many loose threads and so many knots that the great majority amongst the scientific community could not bring themselves to accept the idea, even if it explained many conundrums.

The quotations here provide a vignette of the period during which the idea was to emerge from the realms of scorned fantasy to become a unifying concept that is the basis of much of modern-day geological science. The drama was played out, not in some parochial theatre, but on a crowded stage with players from all parts of the world, and reminds us, once again, that consensus amongst the majority is at any one time not proof of the validity of anything.

THE FALLACY OF DRIFTING POLES

The dilemma facing geologists at the beginning of the twentieth century was an inevitable consequence of the perceived wisdom of the time – that the continents and oceans were permanent fixtures on the surface of the Earth. 'These features of the earth's surface are of old standing . . .' was how Geikie (1903: 24–28) put it in his *Text-book of Geology*. One of the many problems that then arose out of this dogma was the perplexing occurrence of 'abundant [fossil] remains of a vegetation indicative of a warm climate . . . including a bed of coal 25 to 30 feet [7.5 to 9 m] thick' in the Arctic Circle, in latitudes as high as 82° North, where plants could never have flourished on ground permanently frozen or covered for most of the year by thick snow.

This dilemma had spawned various hypotheses suggesting that it was the axis of rotation around which the Earth spins that had changed with time, shifting the North and South Poles to positions very different from where they are today. An increased tilt of the Earth's axis around which it spins while revolving around the Sun was implied, exposing different parts of the globe to either more or less warming than at present. While some, including Lord Kelvin, had played with this idea that the poles might have gradually shifted through 10, 20, 30, 40 or more degrees, others had emphatically put a limit of a few degrees, or fractions of a degree, on movements of this sort. To account for past ice ages, or the distribution of deserts, the doctrine of permanence had therefore focused on the drifting of the poles relative to the positioning of the continents, rather than vice versa.

GEORGE STOW IN SOUTH AFRICA

George Stow (1822–82) was an early, self-taught pioneer who wandered far and wide across southern Africa. In 1870 he championed the idea of a 'simple oscillation of the poles' to explain what he believed had been a progressive change from glacial to subtropical conditions in southern Africa during the time spanned between the deposition of our widespread Karoo Sequence and the Tertiary Period. From the fossils he collected, he concluded that the Algoa Bay area around Port Elizabeth had experienced a climatic change from conditions to be expected today at a latitude of 68° South (for example, the present coast of Antarctica) to those more appropriate to 5° South, close to the Equator. His general thesis – an amelioration of the climate since the deposition of the lowermost Karoo rocks – was thus broadly correct, but his timing of events and his placement of the continents were about as wrong as they could be. He had interpreted the great plains, mountain ranges and deep valleys in the Cape Colony as the aftermath of massive continental glaciation, like that which had overwhelmed northern Europe and America during the Pleistocene Epoch (see Plate 3).

In Stow's eyes, the present-day asymmetric north–south profiles across its mountain ranges, the rounded forms of their cliff faces, and mound-like rocky outcrops that he interpreted as *roches moutonnées*, showed the imprint of widespread ice action (see Plate 4). Talus deposits (unconsolidated waste collecting at the foot of mountain ranges, and choking small valleys) he interpreted as glacial till, and tumbled boulders that had rolled down steep slopes he regarded as glacial erratics dumped there by melted glaciers. In his mind, this was relatively recent glaciation (Pleistocene Epoch). Adolf Schenck, a frequent visitor to this country, soon demolished this idea (1895: 129–41). Schenck recognised that the landforms are in reality due to erosion by water and wind action, with strong structural control by the abundant intrusions of hard, resistant Karoo dolerite.

This lingering concept of the poles of the Earth shifting with time crops up again in Hugh Exton's presidential address to the South African Geological Society in 1898. A further twist to the debate was given by an inventive but rather eccentric observation by E.H.L. Schwarz. In his book, *The Kalahari* (1920: 23), he noted that if the South Pole were to be located in the Indian Ocean, the various occurrences of the Dwyka Tillite (a sediment of Karoo age laid down by widespread Carboniferous glaciation) in India, Australia and South Africa would have been neatly arranged around the pole. The later abandonment of this idea was prompted by the paradox that equivalent glacial deposits in Brazil would then have been formed on the Equator – something that would have needed some nimble footwork to explain.

LAND BRIDGES: IMAGINARY HIGHWAYS ACROSS THE SEAS

Enough was known during the nineteenth century about the southern Gondwanaland continents to recognise that a rational explanation was needed to explain the remarkable similarity between the rock sequences (that is, the stratigraphy) of the Palaeozoic–Mesozoic 'Karoo System' of southern Africa and those of the more distant land masses, and the fossils they yielded (that is, the palaeontological evidence). To serve this purpose, imaginary highways linking the now widely separated continents – land bridges – were invented.

To appreciate the seductive appeal of this concept, we might revisit the pages written by Lyell (1865: 247–73) on what he termed 'The Theory of a Miocene Atlantis'. Here he recounted how Professor Heer of Zurich had in 1855–59 compared the fossil flora of Miocene age (5 to 24 million years ago) in Switzerland with the modern flora of the eastern states of North America, to conclude that the present northern Atlantic Ocean had once been a huge land mass across which the Miocene plants could have migrated westwards, from Europe to America. This had fostered the idea that this mythical continent of 'Miocene Atlantis' had been of a size equal to that of Europe, and that it had sunk beneath the waves in post-Miocene times.

With more than a hint of condescension, Lyell had responded that while such views might have been acceptable to botanists, geologists knew better, and threw his considerable intellectual weight behind the concept of eastward (rather than westward) migration of plant species right across Europe, Asia and then the Bering Straits into western North America, and from there to its eastern side. The irony is that, in this skirmish, neither of the two alternatives had even entertained the possibility of continental drift providing the true answer.

Charles Darwin did not believe in land bridges. Sifting through *On the Origin of Species* (1859), we find damning criticism of the concept of them being 'the wrecks of sunken continents'. Instead of most oceanic islands being 'mere piles of volcanic matter' (1859: 277), he argued, we should find granites, schists and old fossiliferous rocks if these were indeed the surviving peaks of foundered continents. There is, however, no hint that the former attachment of one continent to another might have been an alternative that appealed to him. He was quite adamant about this: 'I do not believe that it will ever be proved that within the recent period continents which are now quite separate have been . . . united with each other' (1859: 278). He argued that not a single mammal was known to be common to Europe, Australia and South America, while plants could have spread very effectively by their seeds being carried by wind, ocean currents, drifting ice, or birds.

Darwin's fundamental doctrine was that identical individuals (species) could never have been fortuitously produced in different regions on Earth by the process he termed 'natural selection' – that each species evolves at a single birthplace and subsequently spreads outwards from there. Adaptation and modification to new environments would then follow. So, how then did he approach the riddle of the present distribution of similar species in North America and Europe? He did this by invoking worldwide climatic change, with warmer climates before the Glacial (Pleistocene) Epoch having allowed species adapted to temperate climates to flourish much closer to the polar regions than would be possible today.

Thus, Darwin argued, Europe, Siberia and North America would at that time have presented a nearly continuous girdle of land across which species could have migrated. The subsequent return of widespread glaciation would then have forced the retreat of species towards less inhospitable climatic zones nearer the Equator. In this way he accounted for the '. . . amount of uniformity in the sub-arctic and northern temperate productions of the Old and New Worlds' (1859: 287) during the late Tertiary Period. He made much of the fact that the various species in Europe and North America showed distinctly stronger similarities during the late Tertiary than they do at present – as he would have expected from the doctrine of 'adapt or perish' that he taught.

The papers by the self-taught Stow (1870), hinting at some form of geographical link between the southern continents, reflect the consensus of these early times. Fossil reptiles, fossil leaves of *Glossopteris* in the Karoo, and molluscs of the Cretaceous and Tertiary deposits near Port Elizabeth, had clearly led him to link South Africa to India and Australia. The link that he imagined was that of land bridges crossing the oceans. The coral reefs in the Indian and Pacific Oceans, he believed, provided evidence that '. . . most of these coral reefs are now crowning the tops of the ancient mountains of a subsidiary continent along the shores of which the various genera of shells and other animals could have migrated' (1870: 548).

Another who embraced this idea was William Atherstone, the medical doctor in Grahamstown who was elected the first honorary Vice-President of the Geological Society of South Africa in the year of its foundation in 1895. In the first issue of its journal *Transactions* he wrote of 'the Zuurberg and Witteberg [Ranges], extending for six hundred miles in three parallel folds through Grahamstown and the Fish River mouth to India . . .' becoming broken when '. . . gradual upheaval of both continents was accompanied by the depression of the bed of the [Indian] ocean between, no trace now remaining of their former union, save in the scattered islands, and in Madagascar . . .' (1895: 76).

The acclaim with which Atherstone's paper was greeted by prominent figures of the Geological Society who attended this inaugural meeting, such as Hugh Exton, FGS (President), A.R. Sawyer (Vice-President) and the mining engineer, Dr David Draper, reveals the general popularity of the concept at that time. Significantly, Exton attributed the early formulation of it to Atherstone himself, claiming that the supporting evidence later brought forward by a number of eminent authorities of the times (such as Professors Bonney, Seeley, Green and Rupert Jones) was '*published subsequent to Dr. Atherstone having formulated his idea* of a pre-existing land connection between the African and Indian continents . . .' (1896, emphasis added).

The size of land bridges envisaged by some can be judged from a remark by R. Lyddeker, Fellow of the Royal Society, to the effect that: '. . . the peculiar lower Gondwana flora in South America, Africa, India and Western Australia . . . leaves little doubt that . . . these widely separated areas were connected by a belt of land embracing the greater part of the earth's circumference in lower latitudes' (1898: 151–53). Not everyone, however, agreed with this extreme view (as clearly revealed in Molyneux's presidential address to the South African Association for the Advancement of Science in 1911).

By 1913, the illusory doctrine of land bridges was so entrenched amongst most palaeontologists that they were even squabbling over whether species migration between South America and North Africa had taken place via the 'Mid-Atlantic' or the 'South Atlantic' routes (for example, Maury [1913], on early Tertiary fossil shells). The concept of bridges was taking on the complexity of a railroad map. A 'Miocene Atlantis' was placed by some in the North Atlantic. Atherstone's ideas became endorsed by Schwarz's belief (1914a) in the 'lost land of Agulhas' – a vanished continent between Madagascar and India. Controversy had reached the stage where it was focused more on where the hypothetical land bridges might have been than on whether or not they had ever existed at all.

A ZOOLOGIST'S MISGIVINGS

In 1922, John Hewitt, Director of the Albany Museum in the Eastern Cape (a post he held for 48 years), published a summary of such zoological evidence as might conceivably throw light on connections between South America and Africa. Without entering directly into the controversy of land bridges as opposed to continental drifting, he highlighted some conflicts between zoologists.

Some argued for the spreading of animal species due to some form of geographical link between the two continents, basing this on the distribution of fish, arthropods and mammals. Others explained the spread in terms of a patchy but purely fortuitous survival of some pre-Tertiary cosmopolitan groups (groups that had originally had worldwide distribution) or insisted on late-Tertiary immigration into South America of groups that had originated in North America. Hewitt pointed out, for example, that certain African birds, the *Trogons*, revealed greater affinities with those in the Orient than those in South America. Fossil genera of certain spiders, the *Archaeidae*, appeared to link South America, South Africa and Madagascar with the Baltic, while tortoises linked South Africa with England, Egypt and India. This brief review by Hewitt is thus useful in dispelling the notion that zoological evidence was clear-cut at the time that the concept of continental drift was gaining some degree of credence.

THE NEW HERESY: CONTINENTS ADRIFT

Tribute for the first integrated outline of a theory of continental drifting must go to Alfred Wegener, a German meteorologist, explorer and, in 1924, Professor of meteorology and geophysics at Graz University in Austria. The first (1915) edition of his book, *The Origin of Continents and Oceans*, was revised through four editions until the last in 1929, shortly before his untimely death during an expedition across the frozen wastes of Greenland.

The brilliance and boldness of his contribution, given the status of geological science and the firmly entrenched ideas at that time, make for compelling reading even today. It is true that he made many mistakes, and the mechanism he offered as the driving force for drifting is untenable. Again, his estimated speed of drifting was couched in terms of tens of metres per year – about 500 to 1 000 times faster than the ~5 centimetres we now know it to be. His interpretation of the Mid-Atlantic Ridge between Africa/Europe and the Americas as stretched and thinned continental crust was wrong. Unnecessarily, he included an actual shift of the Earth's polar axis in his mechanism, a phenomenon that alienated many potential supporters of the concept.

Notwithstanding Wegener's errors, his demolition of many older concepts represents a milestone in Earth science. For example, the migration of matching fossil fauna and flora across land bridges to continents now separated by great oceans would require that these bridges should subsequently subside and conveniently disappear beneath the waves. The implausibility of this scenario was revealed by Wegener's argument that the size of a hypothetical bridge linking Africa, South America, Antarctica, Australia and India would have been so enormous that 'there would be no room for the volume of the world's oceans', and therefore sea level 'would rise above that of the whole continental area of the earth' (1929: 13). With regard to Wegener fitting together the South American and African continents from the matching shapes of their coastlines, this had already been commented on much earlier by others,* although it is said that Wegener had discussed the issue with colleagues as early as 1903.

Wegener's contribution faced a howl of criticism, ranging from references (possibly coloured by anti-German sentiment prevalent

* Ayala-Carcedo et al. (2005) have drawn attention to the Spaniard, José de Acosta commenting in 1590 on the similarities between the American and European coastlines. This was discussed again by Francis Bacon in 1620 and in 1658 by Jean de la Placette, who suggested that the similarities could not have been accidental, and attributed them to excavation of the Atlantic Ocean by the Biblical Flood. In passing, we might note that the fit has, with modifications using the edge of the continental shelf rather than the present coastlines as was done by Wegener, been fully verified by modern computer methods, notably that used by Sir Edward Bullard (see Smith, 1972).

after the First World War) to 'fairy tales' to accusations of being non-scientific. Furthermore, he was a meteorologist, to which many geologists had sourly reacted by denigrating his work as an impertinent intrusion on their territory by someone not qualified to speak on the topic. The Cambridge mathematician, Sir Harold Jeffreys, calculated that the forces that Wegener had proposed for driving continents through the basaltic rocks of the sea floor were, quite simply, totally inadequate. The flaws in the model discouraged those who recognised that continental drift was capable of explaining much that was unexplained at that time. Moreover, Wegener had to face formidable opponents such as the influential American petrologist and chemist, Henry S. Washington, who insisted that the igneous rocks on opposite sides of the Atlantic Ocean simply did not match.

A MAN OF VISION

It was then that the South African, Alex du Toit, emerged as the strongest of the few supporters of Wegener's theory. As early as 1921, he took up the cudgels with a bold declaration of faith published by the South African Association for the Advancement of Science. Here he wrote:

> I am advancing in all seriousness the view, revolutionary and heretical as it will appear to orthodox geologists, that Gondwanaland was a much smaller continent than is usually conceived, that its centre lay somewhat further to the south, that the Carboniferous ice-sheet was an almost continuous mass, and that *the land fragments still preserved represent portions of the ancient continent forcibly torn apart, subsequently modified in outline by erosion, deposition, etc., and now separated by vast stretches of ocean* (1921: 126).

Du Toit was not prone to dramatising his own work, but the italic script in which he presented this text reveals the strength of his convictions, even in the face of a sceptical world.

Following the award of a grant from the Carnegie Institute in Washington, Du Toit spent six months in South America in 1923, gathering geological data for drawing comparisons with Africa. Thereafter he published the books *A Geological Comparison of South America with South Africa* (1927a), and *Our Wandering Continents* (1937).

An important strand in the case for uniting the southern continents was the distribution of the plants of the world in Upper Carboniferous and

Figure 11: Alex L. du Toit (1878–1948), hailed by Professor R.A. Daly of Harvard University as 'the world's greatest field geologist'. The biography by T.W. Gevers (1949) lists 151 publications by Du Toit, in the form of books, reports and papers in just about every field of Earth science, while generations of students have cut their professional teeth on the successive editions of his *Geology of South Africa* (1954). His greatest contribution was, perhaps, his evidence in support of the theory of continental drift as early as 1921, in the face of worldwide scepticism at that time. (Portrait scanned from the *Proceedings of the Geological Society of SouthAfrica*, 1949, with permission.)

Permian times. At that time a distinctive assemblage of ferns and seed ferns had existed in the northern hemisphere, but the 'Southern or *Glossopteris* Flora' in the lands of Gondwana was different. As Du Toit (1954: 344) put it, the two showed 'only a limited community of species' while the 'floras of Gondwanaland display many common species as well as genera'. Attributing this to pure chance would have been the weakest of possible arguments, especially with distant India located so far north in the present day. Du Toit insisted that it was a difference in climate at that time that was the decisive factor in the evolution of the southern *Glossopteris* flora, and elimination of most of the northern types from the southern region.

The pillars of Du Toit's model were the geological and biological unity of the Gondwana continents during the Palaeozoic; their fragmentation and dispersal during the Mesozoic; and the close relationship between South America and southern Africa as revealed by the distribution of Devonian and Permo-Triassic sediments, and the Carboniferous glacial rocks (tillites). He also accepted the explanation of the far-thinking Arthur Holmes, in Britain, that the slow movement of deep convection currents in the heated rocks of the Earth's mantle was the driving mechanism for drifting. Only faith that his analysis was correct could have led Du Toit to take such a bold stand, at a time when another 30 to 40 years would need to pass before most northern hemisphere geologists accepted its validity. In this regard, Ayala-Carcedo et al. (2005) comment that despite Wegener's book having been translated into English, French, Spanish and Russian in 1924, the only supporters of continental drift at an international symposium in the USA in 1926 were the organiser of the symposium and the originators of the theory. As one observer has commented, caustically, 'Their attitude was one of "Hear no Evil; Speak no Evil; See no Evil"'.

In *Our Wandering Continents* Du Toit emerges as a doughty fighter who did not hesitate to challenge the 'big names' of the time (1937: 37ff.). He dismissed the American, R.T. Chamberlin's, attack on Wegener almost contemptuously as 'too full of superficial statements or sweeping generalisations and misunderstandings . . . to necessitate detailed reply'. He took Schuchert of Yale University to task for clinging to the concept of land bridges. Even the influential Californian seismologist, Beno Gutenberg, did not escape criticism for his belief in the loss of much of the Earth's crust to form the Moon, leaving all land concentrated around the South Pole. How then, Du Toit demanded of him, were the continents of the northern hemisphere to be accounted for?

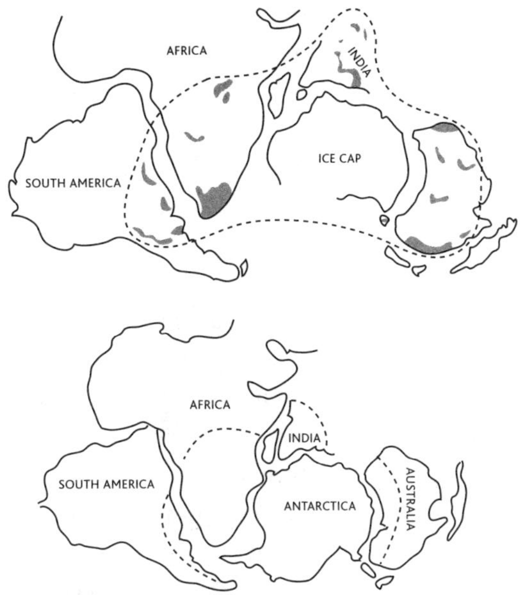

Figure 12: Reconstruction (upper map) of the great southern continent, Gondwana, by Alex du Toit in 1937, based on the information available at that date. From the remnants of the glacial sediments laid down during the Dwyka glaciation (grey areas) and the directions of ice movement indicated by the scratches on the glacial pavements, the map he produced was remarkably close to modern-day reconstructions (lower map) compiled with the aid of computers to find the best fits along the edges of the continental shelves, modern palaeomagnetic data, and more complete knowledge of the geology of the separate continents. The broken line shows the limits of the great ice cap. Readers should note that the scales of the two maps are slightly different.

The concept of continents drifting on the surface of the Earth had not been entirely taboo in the early years of the twentieth century. In 1879 Sir George Darwin put forward the idea that the Moon was torn from the Earth, with the Pacific Ocean representing the huge scar left behind. Much later, Sir Lewis Fermor (1938: 79) laid particular stress on calculations by W.H. Pickering, in 1907, that the density of the Moon (3.4 gm/cm^3) was about the same as that of the average Earth's crust to a depth of around 60 kilometres, and added that this 'certainly adds verisimilitude to the whole lunar hypothesis'. Fermor then interpreted Du Toit's evidence for continental drift in terms of a partial healing of the Pacific scar by the drifting of continents over it. This, he suggested, was accomplished by the break-up of the supercontinent of Pangaea (the greater supercontinent that united Gondwanaland with the northern continents) along a massive rift that opened up as the Atlantic Ocean. As Fermor saw it, 'the loss of three-quarters of the crust [to the Moon] provided the elbow-room for the subsequent break up of Gondwanaland and Laurasia with [the] formation of the Atlantic Ocean'.

MISGUIDED REARGUARD SKIRMISHES

The years of scepticism and denial of drift are implicit in the comments by Jeffreys and Gevers, cited at the start of this chapter, and a host of others. 'It might be thought,' wrote Hallam (1989: 160), 'that the combined efforts of Holmes and Du Toit . . . might have at least begun to sway opinion towards Wegener by the time of the Second World War . . . On the contrary, opinion seems to have hardened against the notion of drift.' The objections were focused mainly on the seeming lack of any mechanism to drive the process, rather than on the positive geological evidence presented. Notwithstanding this, a growing number began perhaps to feel that continued denial of continental drift might leave them in the same category as those who admitted to membership of the Flat Earth Society. But, as late as the close of the 1950s, some rearguard skirmishes were still being mounted. One of the better known was that by Charles Hapgood and James Campbell (1959) in the USA, published as a full-length book, *Earth's Shifting Crust*. A brief commentary on this is not out of place here, as an illustration of the obstinate survival of outdated ideas late into the twentieth century.

Arguments were resurrected in support of land bridges, sunken continents (Nearctis, North Atlantis, Gondwanaland, etc.) and, foremost of all, the permanence of the relative positions of present-day continents. Shorn of all detail, Hapgood and Campbell's hypothesis was one of slippage of the entire crust of the Earth, as a single coherent shell or skin, over the underlying mantle, at intervals of around 40 000 years during the Pleistocene and, by extension, back into the distant past.

This, therefore, was argument for periodic shifting of the continents relative to the Earth's poles, but not with respect to each other – a novel form of continental drift. Hapgood and Campbell's mechanism called for the asymmetric loading of ice on the continent of Antarctica, to produce an unbalanced centrifugal force as the Earth rotated. They claimed that this small but sustained force would be adequate to displace the entire crust along the meridian 96° East. With the circumference of the Earth being some 125 kilometres greater around the Equator than through the poles, they maintained that the shifting crust would come under tension in those regions pushed closer to the Equator, and under compression where crust was displaced towards the polar regions.

As Hapgood and Campbell forcefully remind us in the Preface to their book, Albert Einstein had endorsed their idea in the words, 'I think that this rather astonishing, even fascinating, idea deserves the serious attention of anyone who concerns himself with the theory of the earth's development.' We might, however, suspect from this that Einstein had not kept abreast of current developments in Earth science.

From their basic premise, Hapgood and Campbell theorised that the last position of the North Pole had been in Hudson Bay, then between Greenland and Iceland before that, and close to the Alaska peninsula prior to that in turn, 130 000 years ago. Their explanations for ice ages, mountain building, changing sea levels and extinctions of species then followed from this.

It is not difficult to pick out the many fatal flaws in their presentation, which, it should be remembered, was launched only a few decades ago. Their model was based almost entirely on what they believed had happened during the Pleistocene Epoch, a woefully tiny stretch of time within the history of the Earth. To extrapolate their reconstructions far into the past was mere recklessness. Their rejection of pertinent geological data was quite arbitrary. Although Du Toit's comparison of the geology of South America and Africa appears in their list of 465 references, there is little evidence that they had actually read his book, which they dismissed as based on no more than an 'apparent correspondence in shape between certain continents' (1959: 28). Wegener's work received equally short shrift in their dismissive comment: 'It is interesting to note that despite the quite overwhelming character of objections, attempts are still made to rehabilitate or rescue the Wegener theory' (1959: 30). In spite of their protest that previous ideas on the speed of continental drift had adopted excessive values, they did not appear to regard their own rates of periodic shifting of the inferred positions of the geographic poles by around 3 250 kilometres in 10 000 years (in excess of 0.3 km per year) as unrealistic.

TOWARDS A MODERN PARADIGM: PLATE TECTONICS

The cracks in the wall of opposition to continental drifting began to widen with the contributions by S.K. Runcorn in 1959, on palaeomagnetic evidence. Runcorn was a graduate of Cambridge, and, later, Professor of physics at the Universities of Durham and Newcastle-upon-Tyne, but it was his work on the magnetic properties of rocks that focused the spotlight on him for proving that continents did indeed drift. The apparent shifting of the positions of the magnetic poles in the past, permanently recorded in the rocks of the continents, was shown to be consistent with the drifting

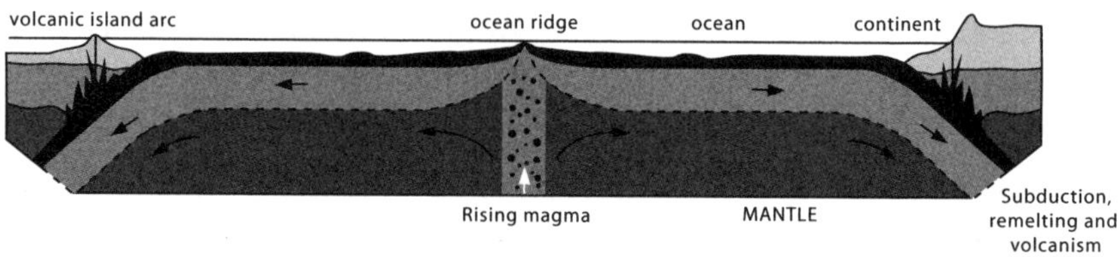

Figure 13: This figure conveys the essence of the Plate Tectonic Model generally accepted today, the basic concepts of which had been proposed as far back as 1928 by Arthur Holmes in Britain. Slow convection and melting in the Earth's viscous *mantle* (dark grey) leads to the emplacement of magma into the *ocean ridges*, where its crystallisation produces the basaltic rocks of the sea floor. As the basaltic rocks (shown in black) cool through what is known as the Curie temperature, the orientation of the Earth's magnetic field at that time is permanently recorded in them. As the process continues, the sea floor becomes progressively displaced to either side of the ridge (see arrows) until it encounters either a *continental mass* or an *island arc* (light grey). Here the sea-floor rocks and the uppermost part of the underlying mantle (the *lithosphere*: medium grey) pass (are *subducted*) beneath the lighter continental or island arc rocks. Melting of the subducted material at depth then gives rise to intrusions of magma into the continental and arc rocks, and the extrusion of volcanic rocks. Earthquakes are generated at various depths along the *subduction zones*, by the stresses developed there.

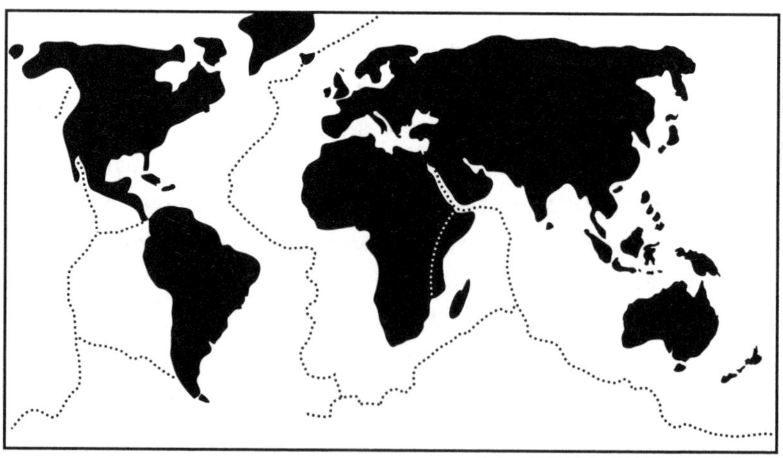

Figure 14: A simplified map of the global system of 'mid-ocean' ridges, forming what has been called the 'longest mountain chain on Earth'.

of the land masses themselves, rather than of the magnetic poles.

The interpretation of mapped patterns of distinctively magnetised stripes (lying more or less parallel with the ocean ridges) in the rocks of the sea floor, and their correlation with specific rock ages by F. Vine and D.H. Matthews at Cambridge during 1959–63, was a vital key in developing the theory of Plate Tectonics – the modern, more comprehensive model for continental drifting. This overcame the hurdle of driving the continents through the strong basaltic rocks of the sea floor, by recognising that the more rigid outermost 100 kilometres of Earth – the *lithosphere*, as distinct from the crust alone – were composed of rigid plates that moved over a deeper, softer *asthenosphere*, carrying the continents with them like ice floes drifting on water. A ploughing of the continents through the rigid, shallow rocks of the basaltic sea floor was no longer an issue.

A NEW BLOSSOMING OF CONTROVERSY

There will always be reaction to new theories from some quarters, so it should be no surprise that the Plate Tectonic Model has had to face a concerted challenge in the opening years of the twenty-first century. At the 'New Concepts in Global Tectonics' meeting held in Italy in 2004, the confrontational mood of thirty-seven participants from fourteen different countries led them to insist that 'plate tectonics is unable to explain the substantial volume of new facts which have accumulated in recent years . . .' (Wezel, 2005).

The Plate Tectonic Model

It has been my purpose here merely to highlight some of the more vigorous controversies of earlier days, so there is no need to launch into a detailed account of the Plate Tectonic Model, which is now in any event widely taught as part of the curricula of most schools. The briefest possible summary of the salient points will suffice. The model tells that:

- the continents are ancient, parts of some being at least 4 400 million years old, whereas the ocean floors are all relatively young, less than 200 million years in age;
- this difference in age is brought about by continuous creation of new ocean floor along mountain-like features – the ocean ridges – which rise from it. This process results from slow convection (bodily movement driven by heat) of hot rock within the Earth's mantle beneath the rigid outermost lithosphere, the separation of magma during partial melting as the convecting plume approaches shallower depths, and the emplacement of rocks of basaltic composition at the crest of the ridge;
- the Earth's magnetic field is subject to periodic reversal, when magnetic north changes to become magnetic south, and, as they cool, the basaltic rocks record the orientation of the Earth's magnetic field at the time of their emplacement. The sea floor thus shows patterns of magnetic 'stripes' of different age broadly parallel with the ridges, by which its parts may be dated;
- as new sea floor is created, the older parts of it, distant from the ridges, slide or are pulled away to either side of the ridges. To accommodate this creation of new sea floor, old ocean floor is forced down back into the mantle along zones termed subduction zones. Deep ocean trenches are created there, and earthquakes and volcanoes become active hazards adjacent to the trenches;
- the pattern of convection in the mantle is thus responsible for the drifting of the continents.

The objections raised by the speakers covered a wide front. It was asserted that the concept of the continents and lithosphere being carried in piggyback fashion on the Earth's mantle, during continental drifting, must remain no more than an unproved idea until a deep drill-hole actually shows that this is so. The permanence of the relative positions of the continents on the globe was once again asserted, with their present-day separation explained by a model of an Earth that is steadily growing larger. Much was made of the fact that mountains are not confined to the margins of continents along fronts where the sea floor is thrusting against them, but are also found in the interiors of continents. The Earth's mantle was claimed to be a cold and rigid solid, while the core was pictured as a low-temperature material which is, atom-by-atom, increasing the volume of the mantle and hence the radius of the Earth.

These concepts all have their roots deep in twentieth century and earlier perspectives, and have raised the blood pressures of some critics in much the same way as the idea of continental drift did 90 years earlier. Only time will tell what the shelf-life of this revival of controversy will be.

7
The Mohole Project: A hole in the bottom of the sea

'A number have warned against drilling into the earth at all, suggesting that if the crust is pierced our globe will pop like a toy balloon or that a hole in the ocean floor will permit all the water to drain out . . . Others suggested that a tiny hole in the crust may set off a full scale volcano.'

Willard Bascom (1961: 67)

The Mohole Project was an ambitious plan aimed at drilling through the Earth's outer crust to intersect rocks of the mantle beneath it. The project was conceived by the American Miscellaneous (AM) Society – a group of scientists who were in the habit of getting together as much for a laugh as for anything else. The proposal rapidly graduated beyond the status of a joke, and by the beginning of 1958 the AM Society found itself with a project director, Willard Bascom, and funding that enabled it to complete an initial feasibility study on a budget of around US$1.5 million.

A series of five test holes was drilled off the west coast of California in 1961, the deepest of which penetrated more than 1 000 feet (300 m) of sea-floor sediment. A further five, drilled at a water-depth of 3 565 metres off Guadalupe Island, penetrated 170 metres of deep-sea oozes before intersecting basalt. This indicated that deep-sea drilling from a ship was certainly feasible, but the ambitious plans for the next phase ran into huge budgetary problems not untainted by political overtones. The project was abandoned when the estimates of the costs of future work reached $75 million and the US Congress withdrew its support in favour of space exploration.

Since that time, drilling of the ocean floor has become an important theme in Earth science, although the depths sought are not as ambitious as those hoped for in the Mohole Project. The *Glomar Challenger*, a US research vessel capable of drilling holes at water-depths of up to 6 000 metres, started recovering cores in 1968. One of the earlier results was the demonstration that the ages of the lowermost sediments that rest upon the basaltic rocks of the sea floor increase steadily with distance from the Mid-Atlantic Ridge. At a distance of some 1 600 kilometres, the age of these sediments was found to be approximately 80 million years, indicating that the distance from points on the sea floor to the crest of the ridge in the south Atlantic has been increasing by some 2 centimetres per year. This provided further proof that the concept of sea-floor spreading, and renewal of the ocean floor, is valid.

The anecdotes from Bascom's book, quoted at the start of this chapter, illustrate the fears still held by some sections of the populace in the twentieth century – that the Earth will pop like a pricked balloon, or that the oceans will drain away if the crust is pierced by a hole or shaft of some sort. If we need any reassurance that this will not happen, we can point to the deep boreholes that were drilled in the Kola Peninsula by Russians during the years 1970–97 (Popov et al., 1999). One of the holes penetrated to a depth of 12.062 kilometres, and the deepest to 12.262 kilometres. There have been no reports of ominous rumblings or gurglings, nor other dire consequences, following this astonishing feat of engineering.

8
FUTURE CLIMATES: GLOBAL WARMING OR ANOTHER ICE AGE?

'General agreement exists, based on tide-gauge measurements, that sea level has risen since the end of the last century by about 1 mm per year . . . Much uncertainty exists about future changes.'
Anthony Hallam (1992: 155)

'Correctly assessing recent global changes in temperature is a very complicated task.'
Brian J. Skinner and Stephen C. Porter (1999: 521)

'There is, in Earth System Science, a growing infatuation with ever more complex models. It's gotten to the point where some models look as inscrutable as nature itself.'
John Harte (2002)

Although geological science is largely concerned with the distant past, it has also become increasingly relevant, particularly within the last few decades, to an understanding of the natural hazards that all of us face during our lifetimes. Within its ambit lies, furthermore, the assessing of the impact of disasters we might have to confront at any time in the future – earthquakes, tsunami, volcanic eruptions, or collision with giant meteorites, to name a few. Events within the last few decades have bonded together geological science, meteorology and physics to address another of the 'hottest' topics at the start of the twenty-first century – global warming. The issue here is whether or not global warming is a transient phenomenon, and what the effects might be if present trends are sustained through the decades and centuries to come.

The questions asked cover a wide spectrum, ranging from the effects of warming on an Earth-wide scale, or on specific countries, down to its effects on entire communities, and from there, ultimately on individual beings. Could climatic changes pose the threat of widespread famine? Could the insidious warming of our environment lead to global rises in sea levels, causing flooding of all low-lying terrains and islands, and become a looming threat to our harbours, bridges, beaches and holiday resorts, or any form of coastal development? Will retaining walls be needed to keep waves out of the reception areas of beachside hotels? Will you be wearing gumboots to paddle across the floor of your seaside cottage? Will the comfortable seat on the rock overlooking your favourite fishing spot be left high and dry, or disappear under the water in the near future?

PREDICTIONS BASED ON LONG-TERM TRENDS

In the *New Scientist* of June 1975, J. Gribbin identified a fundamental control of climate over very long periods of time (1975: 695). He pointed to a cyclicity of major ice ages in the geological record, defining a periodicity of approximately 250 million years. Gribbin noted that W.H. McCrea, an astronomer at Sussex University, had pointed out that this was just about half the time that it takes for our solar system to make one orbit around the Milky Way galaxy, during which the solar system crosses a 'dust lane' every 250 million years. The effect of this would be a warming of the surface of the Sun as it accretes dust. Paradoxically, it was argued that the increased heat output would lead to increased precipitation and cloud cover over the Earth, with increased albedo (radiation reflected) leading to more heat being reflected away from the planet than before, thereby initiating another ice age. Perhaps significantly, it was established that our solar system emerged from just such a situation 10 000 years ago – following our emergence from the most recent widespread glaciation.

Professor L.A. Frakes in Australia (1979) compiled an often-quoted curve showing mean (averaged) global temperatures through geological time. What this showed was three broad cycles of frigidity – Precambrian, Permo-Carboniferous, and Pleistocene, spaced roughly 270 million years apart. The geological evidence for these icy times can be mustered only in a rather general way, as the drifting of continents would have positioned them at different distances from the polar regions as time passed. Glacial phenomena would not, therefore, necessarily have peaked at the same time at all places affected on Earth. If we should extend the last part of Frakes's curve, or that of a more recent compilation offered by McCarthy and Rubidge (2005: 260, simplified here from their data as Figure 15a) into the future, we might conclude that, on a long time scale, we are heading for another glacial stage as an extension of the Pleistocene. By contrast, however, if curves showing the best estimates for the last 18 000 years alone (for example, Coe et al., 2003: 37, simplified here from their data as Figure 15b) were to be projected into the future, we might be persuaded that the Earth is warming up.

Geochemical data from the geological record indicate that the Middle Cretaceous,

Figure 15a

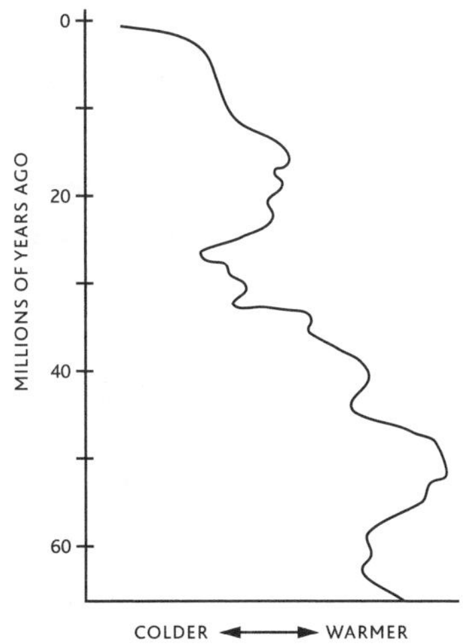

Figure 15b

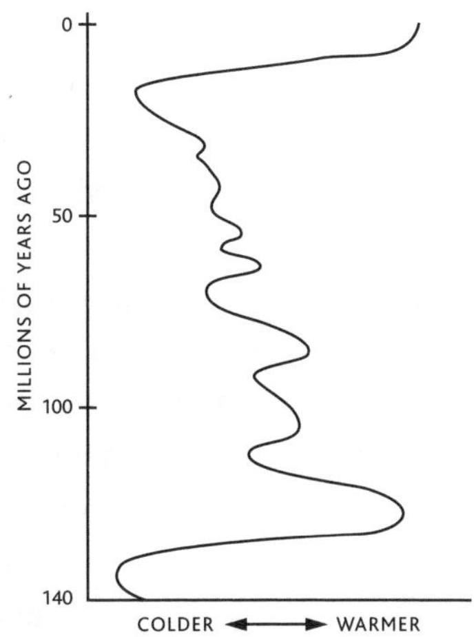

Depending on the time spanned, opposed conclusions might be drawn from trends of temperature variations in the past. Curves show the broad trends of ocean temperatures through (a) the last 60 million years (simplified from McCarthy and Rubidge, 2005: 260) revealing a prolonged period of global cooling, and (b) the last 140 thousand years (simplified from Coe et al., 2003: 37) showing the shorter-term inflections towards higher temperatures during interglacial stages. Curves showing still finer detail reveal global warming between eighteen and ten thousand years ago, with ill-defined cooling thereafter. These data illustrate the uncertainties inherent in using past trends to predict future changes.

the Eocene and the Pliocene were times of markedly higher temperatures than those that prevail at present, with some Cretaceous ocean waters perhaps 15 to 20 °C warmer, and the global average around 8 °C higher (Skinner & Porter, 1999: 525). These periods of catastrophically high temperatures are attributed to high levels of carbon dioxide in the atmosphere, coupled with more active volcanism. If temperatures were again to rise to these levels, the effects on present-day life could well be disastrous.

PREDICTIONS BASED ON SHORTER-TERM TRENDS

As indicated above, long-term periodicity in climate trends helps little in predicting changes on a very much shorter timescale. What interests us the most is what is likely within the next 30 to 100 years, for predictions of what might happen in the next 1 000 or 1 million years are unlikely to cause us sleepless nights.

As recently as March 1975, the 'Monitor' section of *New Scientist* offered the view on page 621 that: '. . . we appear to have suffered, over the past three decades, a retrogression of about one sixth of the way back to that frigid state [of the last ice age]'. The leader page of *The Spectator* of 9 August 2003, attributes to the editor of *New Scientist* at that same time (1975) a warning that 'the threat of a new ice age must stand alongside nuclear war as a likely source of wholesale death and misery for mankind'.

These forecasts were made a mere 30 years ago and yet they stand in stark contrast to what appears to be the global picture at the present time – one of record-high temperatures in parts of Europe, polar-ice melting, and prolonged drought in Africa. Hurricanes such as *Katrina* have left much of New Orleans in ruins, and predictions are that they will become more frequent. Exceptionally heavy monsoon rains have drenched Asia. Some measurements indicate an annual average rise of the sea level by approximately 1.5 millimetres per year, attributable largely to the melting of polar ice coupled with the thermal expansion of water as ocean temperatures rise. Other projections anticipate a rise of the sea level of 0.1 to 0.9 metres within the next 100 years (Huthnance, 2003). On this scale, the good news would seem to be that the present generation of 30-somethings is unlikely to suffer wholesale trauma and disaster, but over a longer period our descendants might well face alarming scenarios.

VISIONS OF CATASTROPHE

Those who have made the calculations warn that the complete melting of the Antarctic ice cap would raise ocean levels by as much as 60 to 80 metres (Coe et al., 2003: 101). It does not take much imagination to picture how even a partial melting of the Antarctic ice would affect any coastal town or village. Picture yourself taking a stroll from the present beach of some coastal city in South Africa, and then turning inland until you have climbed about 40 metres. Imagine the sea to be lapping at your feet. What would you see?

If you have any faith in humanity's ability to adapt to trauma on this scale, the scene should at least fall short of total disaster, for the invasion by the sea would not have been so rapid as to overwhelm the city between one year and the next. Nevertheless, the orderly layout designed by earlier city planners would have been disrupted beyond repair as the advancing sea pushed the city into adopting one stratagem after another. A new city would be rising on what might be perceived as safe ground behind where you are standing, but lower-lying suburbs would have disappeared, as would the freeways and the original harbour piers and installations.

For commerce to have survived, the docks would have repeatedly been rebuilt at higher levels, but with channels to the open sea now clearly marked by buoys to avoid wrecking vessels on submerged buildings. The surf breaking on the shore would be a churning, dirty-brown scum of wood, plastic, uprooted vegetation and litter swilled out of the abandoned factories, hotels, houses and garages. Hordes of gulls would be picking through the debris.

Services in the city would be under unremitting stress. Power failures would be an ever-present menace as cables short-circuited in the salty water, while business efficiency would plummet as service delivery fell into chaos. Sewerage systems would no longer function effectively as the sewers flooded and water backed up into pipes. The spectre of disease would lurk against a background of rising temperatures. The city would be infested with rats driven from previous habitats, and insect pests would multiply out of control. The costs of destruction of prime real estate would run into the billions.

Smaller towns and villages, lacking the resources to cope with all of this, would simply be abandoned.

THE INDICATORS OF CLIMATE CHANGE

Projections into the future are dependent on the accuracy of current assessments of climate change, but here the caveat is that there is no real agreement on what the trend is. One approach has been to study permafrost temperatures, *permafrost zones* being defined as those parts of the Earth where rock or any other Earth materials are below 0 °C for two years or more, a condition that applies to some 24 per cent of the exposed land area of the northern hemisphere. These measurements are regarded as sensitive indicators of change – more so than records of rapidly fluctuating air temperatures. A general warming of the Alaskan Arctic region by 2 to 4 °C over the last 100 years has been established (Romanovsky et al., 2002), with permafrost degradation also apparent in Canada, Russia, China and Mongolia, with the most significant temperature rise being that since the early 1990s.

Evidence for increasing numbers of icebergs in the oceans around Antarctica, and the calving from the Ross Shelf of the largest iceberg ever observed (295 x 37 km), raise the question whether this is another marker that confirms a trend towards a warmer climate. This view was questioned by Long et al. (2002), who suggested that this was largely an artefact of improved modern data-gathering techniques employing radar. However, a review by Showstack (2002) noted that others took the opposite view, and regarded the loss of 5 700 square kilometres of the ice shelf on the eastern side of the Antarctic peninsula within the preceding five years as firm evidence for regional climatic warming. This is ominously close to fulfilment of the prediction made in 1975 by Columbia University scientist Wallace Broecker, who argued that the trend of gradual cooling would be reversed in the ensuing ten to twenty years. On the basis of an eighty-year cyclicity, he theorised that the end of a cooling cycle was approaching, and that 'global temperatures would begin a dramatic rise that, by the year 2000, would bring average temperatures beyond the range experienced during the last twenty years'. Now that the new millennium has arrived, these words have a familiar ring.

The warning implicit in the Antarctic data seems to be mirrored in the Arctic. Kukla (2004) of the Lamont-Doherty Earth Observatory in the USA has given a summary of the topics covered at a meeting to assess recent climate changes in the Arctic. His review begins with the arresting words: 'Alarming observations have been reported on recent climatic changes in the Arctic. Sea ice is thinning and shrinking, and its area was the lowest since 1979 when systematic microwave satellite observations began.' An overall decrease of sea-ice cover over the last 30 years was bolstered by a recent acceleration in the rate of loss. Discharge from Siberian rivers has increased. All of this can be expected to affect patterns of ocean-water circulation. The changes on an annual scale agree with the shift expected from the impact of increasing concentrations of greenhouse gases, but speakers at the meeting warned that the changes occurring in the Arctic could become reversed, and therefore that 'projections of future climate developments remain uncertain'.

THE 'GREENHOUSE GAS' FACTOR

There will be few readers who are not familiar with the term 'greenhouse gases' – gases with marked heat-trapping properties. The chief culprits are water vapour, a group of gases produced by natural processes as well as human activities (carbon dioxide, methane, nitrous oxide and ozone), and chlorofluorocarbons (CFCs) manufactured for use as aerosol propellants, and in refrigerating systems.

Air extracted from bubbles trapped in cores taken from old Antarctic and Greenland ice caps shows an unmistakable correlation between changing levels of carbon dioxide and methane, and Pleistocene glacial and interglacial stages through the last 160 000 years (Skinner & Porter, 1999: 336). Commenting on these data, physicist Harte (2002) has written, 'the net effect of surface warming on the planet's very complicated carbon cycle was the release of more CO_2 [carbon dioxide] into the atmosphere. And that, in turn, further enhanced warming . . . [O]ur present general-circulation models . . . are probably underestimating the magnitude of impending global warming.'

While methane has increased in proportion to the increase of human and livestock populations on Earth, the rapid increase of carbon dioxide through the last 150 years, by around 30 per cent above normal interglacial levels, correlates with increasing industrialisation. Some predictions suggest a global rise in average temperature of 3 °C or more in the next 100 years. In the light of estimates of the average temperature rise in the last interglacial stage having been a mere 1 to 2 °C, this is ominous. Even more sobering is the warning conveyed in Harvey Brooks's *Technology-related*

Catastrophes: 'If all the available fossil fuel resources in the world were consumed, the ultimate increase in atmospheric carbon dioxide concentration would be between five and eight times pre-industrial levels, and there is little question that the resulting environmental and climatic effects would be catastrophic . . .' (1985: 124).

Rising global temperatures would lead to changing patterns of rainfall, melting of glaciers, reduction in the volume of sea ice, and thermal expansion of sea water, and hence a rise in sea level. Patterns of ocean currents would change. Some scientists warn that rising temperatures might even adversely affect the reproductive processes of humans and animals, rendering their eggs sterile.

A further probable result would be the release of methane from what are known as gas hydrates. These are ice-like solids within which methane is locked up in water and sediments of the deeper parts of the sea floor, and beneath regions of permafrost. One estimate points to more carbon being locked up within these hydrates than is presently held in all the coal, oil and gas reserves on land (Skinner & Porter, 1999: 525); other estimates imply that 'more' means 'more than twice the amount' (Coe et al., 2003: 30). Any release of gas from these hydrates would thus accelerate the further accumulation of greenhouse agents in the atmosphere, promoting in turn further warming, and thereby starting a continuous, self-sustaining cycle.

One way to reduce the emission of carbon dioxide would be the increased harnessing of nuclear plants to generate electricity. Dawson (2003) has commented on a recent study at the Massachusetts Institute of Technology that predicts the emission of hundreds of billions of tonnes of carbon as carbon dioxide in the next 50 years unless patterns of energy production and use change drastically. The estimate is that the release of 1.8 billion tonnes of this each year, from coal-fired plants, could be prevented by greater reliance on nuclear power stations, but the caveat here is that this could not be achieved entirely free from consequent problems. The costs of generating electricity from nuclear plants were calculated to be in the region of 6.7 US cents per kilowatt-hour, compared with 4.2 cents for coal and 3.8 to 5.6 cents using natural gas. A second critical issue is that of public safety. Dawson reminds us that public confidence has been severely eroded by the nuclear disasters of Three Mile Island in the USA and Chernobyl in Russia. And then, he adds, there remain questions about the disposal of radioactive waste,

'. . . one of the most intractable problems facing the nuclear power industry'. Considering that waste products currently require isolation for a period of some 300 000 years, a heavy shadow still hangs over the nuclear solution.

To summarise, current forecasts lean to a mostly pessimistic outlook. While it might be possible to curtail the effects of some factors in the equation (industrial discharge, deforestation, neglect of wetland areas, burning of carbon-based fuels, exponential rise in the number of motor vehicles, etc.), there are others over which we have no control (water vapour in our atmosphere, volcanic activity, and emissions from the sea floor). All that can be said is that those of us who have passed their middle years are unlikely to be faced by severe adversity, while there should be some hope that there might be factors, not recognised at present, that nature itself will bring to bear to restore a balance in the near future – that the trends may be cyclic. The present trend could reverse itself within a few years, as it has done in the past, but *The Spectator* article (August 2003: 7) referred to above warns, somewhat cynically, that in the meantime 'spokesmen for the global warming lobby will have trousered a few more grand in television appearance fees'.

DROLL REACTIONS TO THE THREAT OF GLOBAL WARMING

The reaction of politicians to the generation of methane in the digestive systems of farm animals has also had a more comical side. The government of New Zealand proposed a tax of 72 cents per annum on every dairy cow or beef animal, and of 9 cents on every sheep. The storm of protest that followed was both inevitable and predictable. Correspondence and leader articles in the June 2003 issues of *The Daily News* of New Plymouth, in the heart of the dairy-farming industry of Taranaki, suggested that it was time to 'tax the politicians for their hot-air emissions, not our farting livestock', and warned politicians not to eat baked beans. Verbal protest against the flatulence tax was followed up by the dispatch of parcels of manure addressed to members of parliament. Protective gloves were issued to postal workers, and headlines in newspapers trumpeted 'Faeces Hit Fan Over Mail Bid', accompanied by warnings of a national uprising.

In many quarters continuous global warming in the near future is taken to be incontrovertible fact. *The Weekly Telegraph* of 17 August 2005 suggested that 'Siestas Will Help Britons Beat Global Warming'. Quoting research by the Met Office, the article predicted that average temperatures in Britain will increase by up to 5 °C by the latter half of this century, with heatwaves becoming frequent and rainfall declining to half the present levels. It advocated that '. . . a longer midday break, possibly including a short snooze . . .' should become an increasingly common feature of everyday life. The issue of 13 September 2005 reported that the Royal Meteorological Society had gone so far as to publish confirmation of global warming trends, based on a gardener's diary that recorded the number of times he had cut his lawn. The conclusion was that global warming had extended the growing season by one month over the last twenty years.

So, have we found consensus over future climate trends? Not a bit of it. *The Weekly Telegraph* of 7 December 2005 presented a very different set of predictions. Measurements for the years 1957, 1981, 1992, 1998 and 2004, reported by the National Environmental Research Council in Britain, show a slowing down of ocean-water circulation in the North Atlantic Ocean. These changes prompted their science and environment editors to present, under the headline: 'UK's Climate Set to Cool Down', the prediction that 'average temperatures in Britain are expected to fall significantly within a decade because the warm Atlantic current that maintains Europe's mild climate has slowed down by 30 per cent'. Quite apart from some scepticism over the choice of the phrase 'Europe's mild climate', this prediction may well generate considerable global heat on its own.

9
The legend of Noah's Flood

'. . . all the fountains of the great deep [were] broken up, and the windows of heaven were opened. And the rain was upon the earth forty days and forty nights . . . And the waters prevailed exceedingly upon the earth; and all the high hills, that were under the whole heaven, were covered . . . and the mountains were covered . . . And the waters prevailed upon the earth an hundred and fifty days.
. . . and after the end of the hundred and fifty days the waters were abated. And the ark rested in the seventh month . . . upon the mountains of Ararat.'
Book of Genesis (7, 8)

'. . . all the Cataracts
Of Heav'n set open on the Earth shall powre
Rain day and night, all fountaines of the Deep
Broke up, shall heave the Ocean to usurp
Beyond all bounds, till inundation rise
Above the highest Hills . . .'
John Milton (1674: Bk XI, lines 820–25)

'Six days and nights
passed, the wind tempest and storm overwhelmed,
on the seventh day in its course, was calmed the storm and all the tempest
which had destroyed like an earthquake
quieted . . .
And the whole of mankind who turned to sin,
like reeds their corpses floated.'
Lines 121–28 of an ancient Assyrian tablet dated by George Smith (1873) as earlier than the seventeenth century BC

'The Flood, even more than Creation itself, also represents the critical test of our approach to the interpretation of Genesis.'
Alan P. Dickin (2002: 26)

'Science should not be mandated by reading religious texts. That's not the way to make hypotheses.'
Ali Aksu, geologist at Memorial University, Newfoundland
(see Schiermeier, 2004)

Is the Biblical Flood myth or reality? These days few geological texts make any reference to the topic. One reason for this is fairly obvious – does the subject of the Biblical Flood even fall within the domain of geology? There are countless examples around the world where the nature of the sedimentary rocks, or the fossils within them, tell us that in the past continents or parts of them were submerged beneath advancing seas. However, all but a relatively small number of these events date to very ancient times long before the appearance of humans, and are thus irrelevant in the present context. More importantly, the durations of such submergences can be measured in terms of hundreds of thousands, millions, or tens of millions of years. In terms of the geological timescale, the inundation of land for a period of only ten-and-a-half months (the seventeenth day of the second month to the first day of the following year, as specified in Genesis, 8: 13) would virtually be a non-event.

Why, then, raise the subject of Noah's Flood here? As we shall see, whether it was a physical event of any geological significance or not, the grip of the Flood myth on the evolution of geological thinking amounted to a stranglehold for hundreds of years.

RELIGIOUS DOCTRINE, ARCHAEOLOGY AND GEOLOGY

The concept of catastrophic floods has lurked in people's minds for several millennia, but Noah's Deluge, specifically, had been absorbed into the fledgling science of geology in the early nineteenth century under the influence in Britain of men such as William Buckland, the first Professor of geology at Oxford University, and other pioneers such as Revd W.D. Conybeare and Adam Sedgwick. To them, the deposits of boulders and unstratified clays (the residues of Pleistocene glaciation) were the proof of 'diluvialism' – a universal flood that bore out the account in Genesis.

Then George Smith, an assistant at the British Museum, shocked the archaeological world when, in 1873, he published a translation of cuneiform text inscribed on fragments of ancient clay tablets recovered 25 years earlier from near Nineveh in Assyria, by the English archaeologist, Austen Henry Layard. Smith had painstakingly reassembled these tablets, piece by piece, gradually revealing the ancient epic poem known as *Gilgamesh*. His unravelling of the ancient text was far from being some routine task dispassionately undertaken by a salaried scholar. According to Ryan and Pitman (2000: 17), Smith's feverish excitement at the unfolding of the story was such that it reached a stage when he 'dashed into the hallway and in front of the other startled workers on the museum staff began to tear off his clothes'.

The tablets, themselves dated to 660 years before the Christian era, were in turn proved to be copies from Babylonian texts of a much earlier period no later than the seventeenth century BC.

One section of the *Gilgamesh* epic describes how the builder of an ark had assembled his family and 'all the beasts of the field and animals of the field', another the coming of the Flood (described in Smith's lines quoted at the start of this chapter) and the destruction of all humanity apart from a select few. Then came the abatement of the waters and, finally, the release of a dove, a swallow and a raven from the ark in a search for dry land. It is true that there are some obvious differences in the narratives (for example, the duration of the Flood is given in the *Gilgamesh* as fourteen days where the biblical account specifies a period closer to a year), but it is clear that this was an early template with an astonishing but undoubted similarity to the story in Genesis.

The Flood Myth, edited by Alan Dundes (1988), contains part of an essay by James George Frazer (1918) that claims that the legend of a universal deluge is to be found throughout large parts of Asia (Palestine, India, Burma, Cochin China, Kamchatka and the Malay peninsula), Europe (Wales, eastern Russia and Lithuania), the Pacific (Philippine islands, Formosa, New Guinea, Australia and New Zealand) and South, central and northern America, including the inhabited regions of Alaska and Greenland. Apparently, it was not found in the mythology of China nor on the African continent; nor has research up to the present time found such traditional narratives amongst the indigenous peoples of southern Africa.

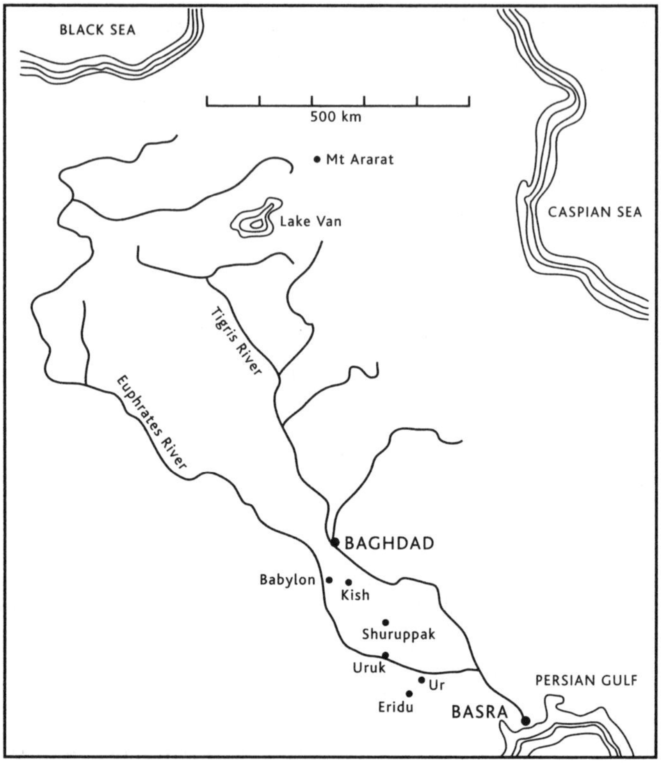

Figure 16: The Mesopotamian region along the Tigris and Euphrates Rivers.

The voluminous early literature on other versions of a universal flood in the Near East, and particularly within the Mesopotamian region between the Tigris and Euphrates Rivers, is amply illustrated by the many contributions and further references in Dundes's compilation (1988). The great antiquity of the tale is left in no doubt, with some scholars tracing it back to the twenty-second century BC. The most persuasive analysis was that by Woolley (1956) who treated the Flood as the history of a definite incident in the time of the Sumerians – a break in the continuity of the history of the country that led to a complete breakdown of the economy, the social system, and rule by the government. According to his interpretation, 'The Earth' is to be read as 'the earth', that immensely fertile accumulation of alluvial silts that had made the establishment and survival of the Sumerian city states possible. The more distant deserts and mountains had, in his view, meant nothing to them, as they were not the fruitful lands that amply supported teeming human life, and were therefore not envisaged as part of 'the earth' of the ancient texts.

Woolley's interpretation appeared to be supported by archaeological evidence. Excavations at Ur during 1928–29 had exhumed eight levels of artefacts revealing progressively older periods of occupation, floored by a substantial layer of water-deposited silt, beneath which in turn were the remains of an antediluvian culture. This he interpreted as evidence for a flood to a height of more than 6 metres that had destroyed lower-lying towns, but had not reached the greater towns built on higher ground. The inundation came at the close of what is termed the Ubaid Period, marking the earliest settlements in Lower Mesopotamia. Allowing for the distortions over time by the repetition of oral narratives, embellishment, and the recopying of ancient texts, we might in this way see the beginnings of the legend.

Perhaps the most elegant explanation for the origin of the story is to be found in a passage from Charles Lyell:

> . . . the catastrophes of modern times might give rise to traditionary narratives, among a rude people, of floods of boundless extent. Nations without written records, and who are indebted for all their knowledge of past events exclusively to oral tradition, are in the habit of confounding in one legend a series of occurrences and incidents which have happened at various epochs . . . (1850: 9).

Lyell was no rabid opponent of orthodox views, but it was inevitable that his rejection of the view that the Earth was a mere 6 000 years old would bring him into conflict with orthodoxy. Not surprisingly, he came in for severe castigation by the clergy and all who supported Creationism (Moore, 1973).

Although he makes no reference to Lyell, Habel (1971) presents another facet of the controversy, showing that the Flood story, as presented in Genesis, was almost certainly compiled from more than one much earlier source. He catalogues inconsistencies in the text of Genesis itself, and differences in literary style in different passages. For example, where one part refers to one pair (male and female) of living creatures to be taken aboard the Ark, another part refers to seven pairs. Similarly, where one part of the text refers to the duration of the flooding extending over 40 days, another part stipulates the lapse of more than a year before fully dry land was once more exposed. Along with other inconsistencies, this is judged to be evidence for a merging into a single account of separate narratives from different earlier versions.

The idea of an Earth entirely covered by water has been extraordinarily persistent in the Western world for two millennia. Seneca (who died in 65 AD), the philosopher and tutor of Nero, harboured images of ceaseless torrents of rain, swollen rivers and rising seas converging to sweep away the entire human race. All of this he wrapped together with woolly ideas on the Earth itself being able to exude living water (*aqua viva*) from the depths, quite unrelated to the water which fell as rain. This vision also appears in Revd Thomas Burnet's *Sacred Theory of the Earth* of 1681, which held that the Sun's rays had split open the Earth's crust and allowed 'the central abyss of waters to burst forth and overwhelm the inhabited lands' (Geikie, 1905: 67).

It comes as something of a shock to learn that John Woodward, a wealthy medical practitioner and ardent fossil collector who endowed a chair of geology at Cambridge University, had some sixteen centuries after Seneca clung to views not greatly different, and taught in 1695 that the interior of the Earth was once filled with water. Disintegration of the Earth's crust had, he claimed, released this water to initiate the Flood (Geikie, 1905: 67). In 1696 William Whiston went so far as to assign a date – 18 November 2349 BC – to the commencement of the deluge and simultaneous release of waters from the Earth's interior following the appearance of a comet.

In his book, *Ever Since Darwin*, Stephen J. Gould, a prolific writer in the USA in the fields of biology and geology, comments that Burnet had assumed 'the truth of scripture and fashioned a physical mechanism to make it happen'. A geological interpretation of Noah's Flood was sought 'not by miracles or divine caprice, but by natural physical processes' (1978: 144). Elsewhere Gould then comments that by the 1840s, the notion of the Flood as a serious geological proposition had passed from the scene. This is not correct. For example, the palaeontologist J.W. Dawson of McGill College, Montreal, in his book written for the Religious Tract Society (*c.*1880), proposed a name 'Palaeocosmic Man' for those whom he identified as antediluvians – 'roving tribes whose remains are known to us [to] represent . . . part of the race of Cain . . .' and that '. . . the catastrophe which destroyed these rude and lawless men was that which culminated in the deluge of Noah, which may represent the extinction of the last great body of the primitive race. . .' (*c.*1880: 245–47).

The stance adopted in this debate by the earliest geologists was coloured by religious convictions, and by the lack of any reliable dating techniques at that time. According to McGowan (2003), William Buckland, the influential Oxford University Professor, remained a committed believer in the Noachian Flood right up to the time of his death in 1856. So did the eminent Frenchman, Georges Cuvier.

Thus it is not surprising that we find a jarring note in the account written in 1857 by W.G. Atherstone (see De Klerk, 1995a, 1995b) of an 1845 expedition with Andrew Geddes Bain that recovered from Cretaceous sediments exposed in the Bushmans River Valley, Eastern Cape Province, what turned out to be the first collection of dinosaur remains in South Africa. Atherstone's words, '. . . we were all engaged overhauling leg bones and hip bones, and diving deep into the mysteries of the world before the Flood' (see De Klerk, 1995a: 28), reveal how deeply ingrained the concept of the global Flood was, even amongst the scientifically minded.*

It was W.J. Sollas, at Oxford University, who had in 1905 called to the attention of the scientific community Eduard Suess's meticulous analysis of the legend that appears in his scholarly book, *Das Antlitz der Erde*, translated into English as *The Face of the Earth* by Hertha Sollas of Newnham College, Cambridge, in 1904. In this, Suess devotes a full 55 pages of text to the subject, analysing line by line many of the ancient Chaldaean legends, reconstructing a path along which the Ark was indicated to have drifted from the mouth of the Euphrates River towards the interior, and testing how various details in the ancient texts might be consistent with the geology of the Mesopotamian region. Asphalt (pitch), for example, which ancient texts state was used in the construction of the Ark, occurs in abundance in hills of Miocene strata overlooking the valley of the Tigris and Euphrates Rivers. Suess rejected the belief that the biblical 'Mountains of Ararat' were those known by that name today.

It was Suess's conclusion that a great wave had been created in

* This is further illustrated in Bain's anecdote, republished (1895: 62) after his death, of an incident in his early years, when he was chiselling from its matrix the skull of a vertebrate fossil that subsequently became known as the 'Blinkwater Monster'. Responding to the questions put to him by an inquisitive young farmer, Bain prudently kept the conversation to simple ideas by referring to the skull as 'the petrified head of a wildebeest'. Warming to this theme, Bain then asked, 'Did you never read that when Noah was in the Ark . . . one of the wildebeests jumped overboard, and before Noah could get out his lifebuoy it was drowned?' A further remark that 'the waters covered the tops of the highest mountains, and at that time Noah was floating above the lofty Winterberg' reveals the immaturity of the concepts of the time. How fossil remains came to be preserved in the eroded, stratified Karoo rocks at the time when, in his view, the present topography was already formed, was not explained by Bain.

the Persian Gulf by the coincidence of a cyclone and an earthquake generated in or south of the Gulf, and that this had swept a great mass of water inland, stranding the Ark on Mt. Nizir. This we would today term a 'tsunami', popularly but inappropriately known as a tidal wave. Suess drew comparisons with the Lisbon earthquake of 1755, when a wave said to have been 60 feet (18 m) in height had killed 60 000 citizens within six minutes, and with cyclones in the Bay of Bengal during the years 1737–1876, one of which had drowned an estimated 215 000 people. Suess also dwelt on the manner in which earthquakes may produce cracks at surface, from which water and mud are explosively jetted, to account for lines in ancient texts describing the 'rising of the waters of the deep' (1904: 25ff.).

W.J. Sollas was highly sceptical of Suess's analysis. The height of Mt. Nizir is some 1 000 feet (300 m) above sea level, which he deemed to be an impossibly great height for any cyclonic surge to have reached. A further significant point, stressed by both Suess and Sollas, is that ancient Egyptian texts, contemporaneous with those of the Chaldaeans, make no reference to any such deluge, thereby implying that the Biblical Flood was a localised catastrophe that never impacted on the Mediterranean region. The final summation by Sollas was vehement: 'Thus the deluge story is a myth within a myth, containing statements clearly unveracious; and how we are to distinguish in this mass of fiction the true from the false passes the wit of man to conceive' (1905: 320).

AN ANSWER IN EGYPTOLOGY AND THE CONSTELLATIONS OF STARS?

An entertaining resurgence of the idea of a Great Flood appears in some recent popular works on Egyptology and other ancient civilisations. In his *Fingerprints of the Gods*, Graham Hancock (1995: 193–99) reminds us that more than 500 legends of a deluge are known around the world, originating in Asia, Europe, Africa, the Americas, Australia and the Pacific. A common thread of many of these legends is that the progenitors of humanity and the animals of today survived only because of the building of a raft or boat which saved them from sinking beneath the rising waters – tales eerily close to the biblical story of Noah and the Ark.

There is little, if any, direct geological evidence that helps in dating the building of the pyramids of the Giza plateau on the banks of the Nile River, but Bauval and Hancock (in *The Keeper of Genesis*, 1996) have interpreted the spatial alignment of the Sphinx and the pyramids, and that of shafts of unknown purpose within the interior of the pyramid of Khafre, in terms of their alignment with specific stars. They emphasise that ancient hieroglyphic texts stress the preoccupation of Egyptian religious rites with the constellations. Bauval and Hancock's thesis is that precession (resembling the wobble of a spinning top) of the axis of rotation of the Earth, with a periodicity of 25 900 years, can be used to fix a date at which the alignment of these features of the Giza necropolis with certain stars would assume unique astronomic significance. This date, on their reckoning, turns out to be 10500 BC, which they propose was the time before the Flood destroyed an ancient but advanced civilisation – one that preceded the completion of the pyramids in approximately 2500 BC.

The ideas that Hancock advances are immensely entertaining, but unfortunately he effectively shoots himself in the foot in attempting a geological explanation for the Flood in terms of a catastrophic event in which the rotation of the Earth is claimed to have been upset by asymmetric loading of the polar regions by ice caps, thus reviving the discredited ideas of Hapgood and Campbell (1959) on continental drift. The result of this catastrophic event would, Hancock (1995) claims, have been a shifting of the Earth's outermost crust 'over the soft inner body, much as the skin of an orange, if it were loose, might shift over the inner part of the orange all in one piece'. Based on this hypothesis, he proposes a model in which 'those parts of the earth's crust which are situated at the North and South Poles (and which are therefore as completely glaciated as is Antarctica today) shift suddenly into warmer latitudes and begin to melt [*sic*] with extraordinary rapidity' (Bauval & Hancock,1995: 468).

Setting aside questions as to whether the Earth as we know it could survive such a cataclysmic event, and the fact that no evidence for a dramatic displacement of the Earth's entire crust is evident in the palaeomagnetic evidence (that is, in the magnetic properties of rocks, which record their geographic latitude at the time of their formation), such a mass movement of the crust would have rendered quite meaningless and fortuitous the precise orientation of the pyramids and the Sphinx that Hancock claims prevails at the present time – a feature he lays stress on. The preservation of the same precise pre-Flood orientation of the ancient monuments of the Giza plateau after such a catastrophic event, capable of shifting polar lands closer to the Equator, would require such enormous coincidence that the entire thesis is blown away.

WHAT THE FOSSIL RECORD TELLS US

If any further rebuttal of the concept of a universal flood is still deemed necessary, we could turn to the essay by Stephen Gould (1982) who took up his pen to demolish, on purely geological grounds, the ideas of fundamentalist Creationists in the USA, following his embroilment in cases before the federal courts that were to decide what should and what should not be taught in state schools. His essay is a rebuttal of the thesis of John Whitcomb, a Professor of Old Testament studies, and Henry Morris, a Professor of hydraulic engineering, who had in 1961 returned to Creationist doctrines to defend the literal text of Genesis.

The world over, the fossil record shows a progression from single-celled life forms in the oldest rocks, to invertebrates in later strata, then the first fishes, followed by the amphibians and reptiles and, finally, mammals. If, Gould asked, all creatures had simultaneously been present on the surface of the Earth, as Creationists would have it, how was this sequence in the rock strata to be explained? Morris had attempted an explanation based on the assertion that the fossil sequence in the rocks was merely a reflection of different rates of settling of animal remains through turbulent waters during the Flood, that is, that some remains were more dense or more streamlined than others, and would therefore sink more swiftly in accordance with the principles of physics. As shown by Gould, this approach is entirely false – the remains of massive and heavier animals are commonly found overlying those of delicate floating creatures that would have settled very slowly through water, turbulent or otherwise.

With regard to the assertion of Whitcomb and Morris that certain animals had, by virtue of superior intelligence or agility, weathered the catastrophe for a longer period of time so as to be entombed only in the highest strata, Gould pointed out that slow-moving and clumsy animals like the sloth are in fact confined to strata that lie above those bearing abundant remains of nimble creatures such as small dinosaurs. Likewise, trilobites, many of which were free swimmers, do not appear in rocks younger than 250 million years in age, but are found in abundance in Cambrian rocks of greater age.

Somewhat surprisingly, Gould did not capitalise on what must surely be the most telling point of all – the duration of the Flood. The Whitcomb-Morris approach, and indeed Creationist doctrine itself, would have the deposition of fossils take place within a single catastrophic event. The duration of this event, in the context of geological time, would have been vanishingly small. Yet the development of precise radiometric methods for fixing the age of rocks shows that the time lapse between the most ancient strata, bearing primitive unicellular fossils, and those containing fossils of early humans, was at least 3 250 million years, and possibly as much as 3 800 million years if we follow the chronology of Nisbet and Sleep (2001).

The fossil record in southern Africa is not as complete as that in many other countries, but there are important aspects within it. It is in the ancient Onverwacht and Fig Tree Groups of rocks in Swaziland that the evidence for the existence of primitive micro-organisms now preserved in sediments is most convincing. These rocks are certainly older than 3 200 million years. Nama Group fossils of the soft-bodied coelenterates of *Ediacaran*-type, generally regarded as of Late Precambrian age, appear in Namibia. Invertebrate trilobites, brachiopods and molluscan bivalves appear in abundance in the Late Ordovician–Devonian Cape Supergroup in the southern Cape. The Upper Palaeozoic Karoo sediments are uniquely important for the vertebrate mammal-like reptiles harvested from them, but contain no trilobites. Hominid fossils, representing primitive humans, appear within cave breccias only a few million years old, at the well-known localities Taung, Sterkfontein, Swartkrans and other sites. This enormous span of time encompassed by the fossil record in this country can in no way be reconciled with Creationist doctrine.

FLOODING OF THE BLACK SEA: A NEW APPROACH

Oceanographic, geological and archaeological research in and around the Black Sea (Ballard, 2001) has hit the headlines in recent years. It has long been known that the Black Sea was originally a landlocked, freshwater lake, with the molluscan fossils in the older sediments of the lake floor being freshwater species. Two thousand years ago the philosopher, Strabo (54 BC–25 AD) had speculated about how the Black Sea and the Mediterranean came to be united. The breaching of an isthmus separating the Sea of Marmora from the inland lake was asserted by Sir Archibald Geikie as far back as 1903, and again briefly presented by Arthur Holmes (1965: 1 194).

It was in 1970 that the Black Sea region became the focus of intense interest following the drilling of a series of holes into its floor by the research vessel, *Atlantis II*, carrying scientists from the Woods Hole Oceanographic Institute in Massachusetts. The story of this voyage is told in *Noah's Flood* by William Ryan and Walter Pitman (2000), geologists with the Lamont-Doherty Earth Observatory of Columbia University, in their fascinating account of how the researching of the Black Sea progressed during the following 30 years.

Geologists on board the *Atlantis II*, David Ross and Egon Degens, described how wherever sea-bottom cores had been recovered, they showed the same transition from a black, stinking, sulphurous and methane-rich mud to an underlying hard, light-coloured clay. The clay was attributed to sediment deposited from fresh water brought in by the Danube, Dnieper, Dniester and Don Rivers during a time when the Black Sea was a vast freshwater lake. The overlying muds pointed to subsequent invasion by salt water, killing off all animal and plant life there. Investigation of the fossil fish, plankton, crustaceans and molluscs revealed that none of the freshwater species had survived the transition – they were abruptly replaced by saltwater species. Old beach deposits had been found at 105 metres below present sea level. The picture was thus one of a sudden overwhelming of the freshwater lake by vast volumes of salt water at a time when the surface of the lake was hundreds of feet lower than at present.

In 1993 Ryan and Pitman participated in a further expedition that was a cooperative venture with Russian scientists aboard their vessel *Aquanaut*, sampling the Black Sea floor offshore from the mouth of the River Don, which enters the Sea of Azov – the north-eastern limb of the Black Sea. This was followed by sampling off other rivers, the Dniester and the Dnieper. Wherever sampling cores were recovered, the same sequence of olive-grey muds overlying hard clay with sandy layers was found. Significantly, the lower layer revealed woody material and fossilised roots, and desiccation cracks filled with dust (like those that can be found today in any dried-up muddy dam or puddle). Sandy layers displayed what is known as aeolian cross bedding, such as can be seen in modern sand dunes. Without question, this had at one time been dry land, and on this cruise the lowest shoreline was traced to 158 to 168 metres below present sea level.

It was the subsequent dating of the fossils that had been recovered that supplied the real excitement. Using the most advanced technique available, known as accelerator mass spectrometry, the age of the interface between the overlying mud and underlying clays was determined. The ages proved to be remarkably consisten – 7 500, 7 580, 7 510, 7 510 and 7 470 calendar years.

Armed with this information, Ryan and Pitman set about tying together a number of threads to formulate a new view of the Flood myth. They used archaeological data; ancient texts such as *Gilgamesh*; oral narratives passed on through the millennia by wandering singers, such as *The Epic Tale of Altraharsis*; blood-typing to trace the movement of populations in prehistory; and linguistic material tracing the common roots of languages. The picture they draw is one with the shores of the Black Sea once populated by thriving communities of farmers, hunter-gatherers and herders. They had crowded the shores when the lake was a vast stretch of fresh water during a particularly dry period (the Younger Dryas Period: 10500–9400 BC) when even Lake Van at the headwaters of the Tigris and Euphrates Rivers had dropped 245 metres to become a salt lake. The surface of the Black Sea had continued to fall by evaporation until at around 7600 BC its level stood 150 metres lower than that of the Mediterranean, from which it was separated by only a narrow barrier, across the Bosphorus Straits.

By now the last ice age had ended, and the melt waters of the continental ice-sheets had been raising sea levels all around the world. Eventually the barrier at the Bosphorus Straits was breached, and the sea poured in. Ryan and Pitman believe that this was an event of unprecedented violence as the barrier was rapidly swept away and a channel was scoured to a depth of 145 metres by the energy of the inflow. Calculations suggested that sea water poured in at a rate of 40 cubic kilometres per day, 'two hundred times what flows over Niagara Falls' (Ryan & Pitman, 2000: 245). As the water thundered in, the plants, molluscs, fish and plankton died in the salty water, and the lake began to rise by six inches (15 cm) and spread its shoreline by a mile (1.6 km) each day. In panic and bewilderment, the inhabitants of the shoreline would have fled.

The reconstruction of this event by Ryan and Pitman is as gripping as anything to be found in the literature of geology: 'It is hard to imagine the terror of those farmers, forced from their fields by an event they could not understand, a force of such incredible violence that it was as if the collected fury of all the gods was being hurled at them.' The authors argue that the dispersal of the displaced communities would have led to waves of emigrants: Semites to the Levant, the Ubaid (forerunners of the Sumerians) to Mesopotamia, and others to the Caucasus, Hungary, the Adriatic, the Aegean, and Russia south of the Urals mountains. Those who settled between the Tigris and Euphrates Rivers would have carried with them, in traditional oral narratives, their memories of the Black Sea event, which would in time have become merged with accounts of the periodic floods that swept the region they had settled in. This, as Ryan and Pitman see it, was the origin of the Flood legend, a story that recalled a catastrophe of 7 500 years ago.

This is a vivid story built on a foundation of good science and impressive scholarship, but there remains one aspect that you as the reader must resolve for yourself. How closely does this reconstruction mesh with the biblical account, the epic of *Gilgamesh*, or other similar legends? Each of these insists on an abatement of the Flood and a return to normality after the peak of the catastrophe. Whether we accept that the submergence lasted a mere seven days or more than a year, the ancient texts specify a dénouement in the subsequent recession of the waters, and this does not fit the history of the Black Sea. The older shores here remain under water to the present day. Ryan and Pitman address this contradiction only obliquely, by asking: 'what if the myth story is examined in a different light, one that asks if human history was cleaved into a before and after primarily because the inundation was permanent and not temporary?' (2000: 49). Their thesis is that the flight of the refugees to the safety of new lands across Asia and Europe would have persuaded them that what they had perceived as a world-embracing Flood had then abated. This belief would in their minds have been confirmed by the limited duration of the many floods that periodically inundated the Mesopotamian region where the Ubaid were now settled.

The controversy about the Black Sea inundation has recently resurfaced again, this time in the journal *Nature* (Schiermeier, 2004). It now appears that geologists and oceanographers cannot agree about the time that it would have taken to top up the ancient lake after the last ice age. Where Ryan and Pitman had estimated this to be three years, leading to rapid inundation of the Black Sea shores, computer modelling is claimed to suggest 33 years as a more realistic timeframe.

Other critics now suggest that the rise in level was not the 100 metres accepted by the earlier researchers, but a much more modest 18 metres. Here is clear conflict between geologists and oceanographers.

NEW EVIDENCE FROM MESOPOTAMIA

The increased precision of rock dating within recent decades, using radiometric methods combined with the counting of varved sediments (layers deposited each year by melting ice) and dendrochronology (a method that relies on the number and characteristic 'banding signatures' of thicker and thinner growth rings of trees), has opened another avenue of search for evidence relating to a flood of global extent. The subject has been thoroughly reviewed by Alan Dickin (2002) in his book *On a Faraway Day*.

Dendrochronology is regarded as accurate back to 9700 BC. The calibration, each one against the other, of carbon-14, ^{234}uranium, ^{230}thorium and dendrochronological methods, and analogous methods using fossil corals, then allows a variety of approaches for determinations of age dating back to around 9000 BC. With further calibration of these methods against sequences of varved sediments in different parts of the world, reliable dating can be pushed back to close on 38 000 years. What emerges from this is that there is no break in the record, and hence that no flood of global extent could have occurred within the last 38 000 years, nor within the last 13 000 years in the source regions of the Tigris and Euphrates Rivers. Dickin concludes in his review that 'interpretations of Genesis which call for such a flood are misguided' (2002: 36).

In short, the story of Noah's Flood as one of global extent cannot be substantiated by any geological evidence: what geological evidence bears on it merely refutes the scale of the event. None of this, however, precludes a local flood, or series of floods, in the Mesopotamian region, which is an entirely different matter. This is the thrust of Dickin's (2002) survey. Whether or not we accept his conviction that Adam and the Patriarchs were all real people is not at issue here: what is important in the present context is his review, as a distinguished Professor in the department of geology at McMaster University in Canada, of the geological evidence. His interpretation is that the Biblical Flood was a real event, but one of localised extent that occurred in prehistory, and that was reinvented in Mesopotamian literature by being 'conflated with more local floods' that occurred several thousands of years later. In this respect, his conclusion is not greatly different to that of Lyell, quoted earlier, but, unlike Lyell's generalisation, Dicken's has a solid foundation of geological data.

As reviewed by Dickin, excavations at the ancient cities of Kish, Shuruppak and Ur have found geological evidence of floods at, respectively, around 2600 BC, 3000 BC, and in times preceding the Ubaid Period (5400–4000 BC). The combination of geological and archaeological evidence points to localised floods at separate times, depositing layers of alluvium that do not correlate at different localities. For example, at Eridu, 'the oldest city in southern Mesopotamia' located only 20 kilometres from Ur, there is no silt layer equivalent to that at Ur. Dickin (2002: 76) therefore argues, on this evidence, that 'we are forced to conclude that the Biblical Flood . . . predates both Eridu and Uruk', and must be treated as a pre-5500 BC event.

Following the worldwide recession of the sea level during the last glacial maximum some 20 000 years ago, the ensuing sea-level rise would, Dickin argues, have impacted on the Mesopotamian region, and particularly so where the Tigris and Euphrates Rivers had reached the stage defined by geomorphologists as that of 'old age', when their lower reaches would have exhibited gentle gradients in what he describes as 'one of the flattest plains on Earth' (2002: 43). The Mesopotamian delta region would thus have been very susceptible to flooding, particularly after an accelerated sea-level rise dating from 8000 BC. Furthermore, lake-floor sediments at Lake Van, between the headwaters of the two rivers, provide a complete record back to 12280 BC, but identify the period 6200–5500 BC as one of exceptional wetness brought about by increased rainfall. This combination of low gradients, sea-level rise and heavy rains is thus postulated to indicate the date of the Flood, at some time between 5550 and 6000 BC.

In the end, it becomes a matter of personal belief whether you accept the Flood story as fact or fiction. Geological science cannot give an answer – it can merely establish what was possible and what was not. All that is certain is that the shadowy legend will continue to linger for countless generations.

PART TWO

Some Early Tangles in Southern Africa

10
Introduction: Confusion over Origins and Ages

Without an appreciation of the concept of time, and the means to measure its passage, we would be unable to make much sense of the geology of southern Africa. Here is a record of bizarre animal and plant fossils, past ice ages, spreading deserts, massive outpourings of lavas, great intrusions of igneous rocks, inundations by the sea, and ancient mountain ranges thrown up and now worn down to stumps. While palaeontology (the study of fossils) was the base from which theories on the evolution of living things grew, it was the distribution of their remains in sedimentary rocks of similar ages that made it possible to prepare the first geological maps. The rub here is that richly fossiliferous rocks, dating from just before the Cambrian Period, represent a span of time that amounts, roughly, to a mere one-eighth of the time elapsed since the formation of the Earth. Furthermore, only relative ages of strata could be determined in this way: absolute ages in terms of years could be no better than educated guesses prior to the early twentieth century. These guesses were both generally and spectacularly wide of the mark.

In southern Africa the problem was the abundance of rocks of great age pre-dating the Cambrian Period, or others that had an igneous origin, or had been recrystallised (metamorphosed) by intense heat or pressure. Those ancient strata that did carry some hint of remains of ancient life yielded

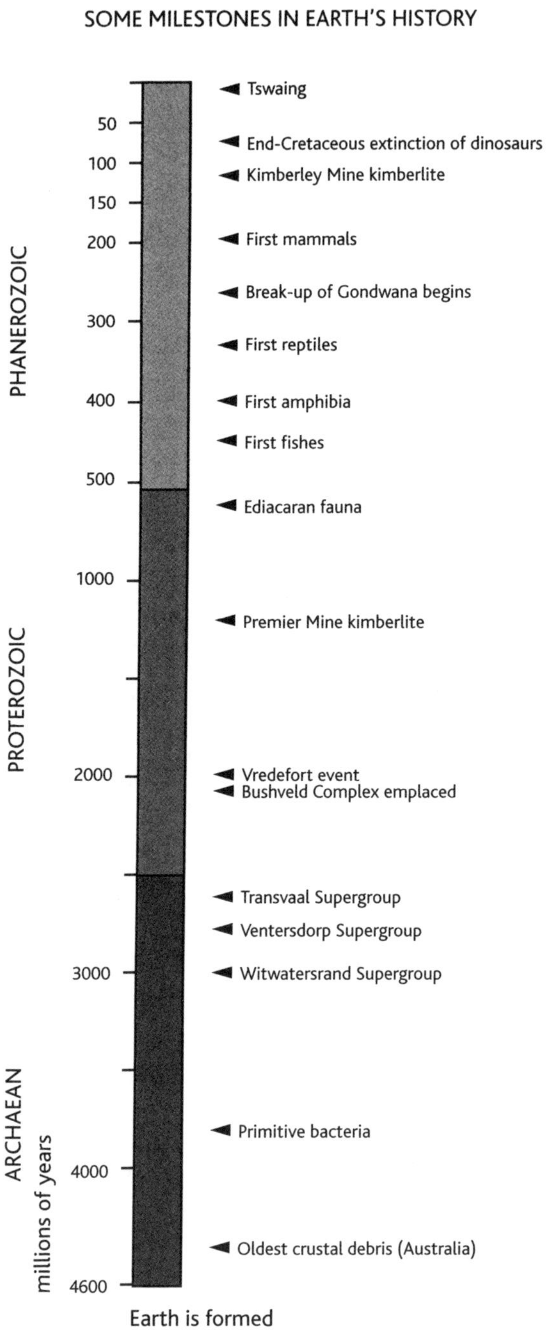

Figure 17: Diagrammatic representation of some of the more significant events in the geological history of southern Africa.

only primitive algae and bacteria. Palaeontology therefore had little application in the great accumulations of rocks that we know today as the Witwatersrand, Transvaal, Ventersdorp and Waterberg Supergroups. It is only in the late Precambrian sequence that we term the Nama Group, and then in the Silurian–Devonian Table Mountain Group of the Cape Supergroup, that fossils begin to appear in some abundance in our record. With the advent of appropriate modern instruments – the mass spectrometer and ion microprobe – the seemingly intractable problem of determining absolute ages was finally overcome.

In Part 2 we examine some of the traps into which South African geologists fell in the years preceding the twentieth century: for example, equating the Carboniferous Dwyka Tillite with the very much younger Pleistocene deposits of Europe, and the confusion that surrounded its origin; as well as the correlation of the ancient Witwatersrand gold-bearing sediments north of the Vaal River with much younger rocks in the Cape Colony (see Plates 5 and 6). The origin of the showpiece of South African geology, the Bushveld Complex, is a topic that has stoked the fires of heated controversy, not only in this country, but overseas as well. Few people today remember the 'Kalahari Scheme' of E.H.L. Schwarz, but at one time it embroiled geologists, farmers, the general public, and

A SIMPLIFIED SEQUENCE OF THE SOUTH AFRICAN ROCK FORMATIONS AND EVENTS REFERRED TO IN THE TEXT

ERA	PERIOD	AGE (million years) AT START	FEATURE or EXAMPLE IN SA	CHARACTER
CENOZOIC	Quaternary	2	Pretoria Saltpan	granite breccias & saline muds
	Tertiary	65	Alexandria Formation	marine limestones & windblown sands
MESOZOIC	Cretaceous	145	Uitenhage, Zululand & related Groups	fossiliferous marine & terrestrial sediments
	Jurassic	210	Drakensberg lavas	basaltic & rhyolitic lavas
			Clarens Formation	desert sands
	Triassic	250	Elliot & Molteno Formations	fluvial sandstones & shales
PALAEOZOIC	Permian	290	Beaufort Group	terrestrial & fluvial sandstones & mudstones
			Ecca Group	sandstones, shales & mudstones
	Carboniferous	350	Dwyka Group	glacial deposits and shales
	Devonian	410	Witteberg Group	deltaic quartzites & shales
			Bokkeveld Group	fossiliferous sandstones & marine shales
	Silurian	440		
	Ordovician	500	Table Mountain Group	quartzites, shales & a glacial bed
	Cambrian	545	Cape Granites	intrusive granites
PROTEROZOIC	Late Proterozoic	900	Nama Group	quartzites, shales & limestones
			Malmesbury Group	slaty rocks prominent
	Middle Proterozoic	2050	Waterberg Group	red-pigmented sediments & lavas
	Early Proterozoic	2500	Bushveld Complex	intrusive basic rocks & granites
ARCHAEAN		3800	Transvaal Supergroup	quartzites, shales, dolomites, lavas & glacial deposits
			Ventersdorp Supergroup	largely volcanic rocks
			Witwatersrand Supergroup	quartzites, shales, conglomerates
			Basement Granitoids	intrusive granitoids of various ages
			Barberton Supergroup	schists, quartzites, lavas

The Cape Supergroup includes the sequence from the Table Mountain to the Witteberg Groups, and the Karoo Supergroup sequence from the Dwyka Group to the Drakensberg Group lavas. Ages, rounded off, are a compromise between those adopted by the S.A. Geological Survey (Hartzer *et al.*,1998) , theUS Geological Survey, and other sources.

ultimately the government in heated argument and threatened to become an international incident before the scheme was defused and forgotten. The age first assigned to the 'Pretoria Saltpan' was subsequently found to be incorrect by more than 2 billion years, and the more modest riddle of the 'Schoenmakerskop quartz porphyry' took many years to solve.

The Cape Supergroup includes the sequence from the Table Mountain to the Witteberg Groups, and the Karoo Supergroup that from the Dwyka Group to the Drakensberg Group lavas. Ages, rounded off, are those adopted by the South African Geological Survey (Hartzer et al., 1998).

11
Claystone porphyry: The enigmatic Dwyka Tillite

[Claystone porphyry], a curious name given to the Dwyka Tillite by *Andrew Geddes Bain (1856)*, and the multifaceted controversy that flared up over the origin of the deposits.

The 'Karoo Sequence' or 'Supergroup' refers to a succession of some 35 000 feet (11 km) of now-lithified muds, silts and sands, capped by basaltic lavas, that spreads over roughly half the area of South Africa, and extends from there into neighbouring territories (see Figure 18). The Dwyka Group is that part of it which lies at the base of the sequence, resting in the southern Cape province upon the underlying Cape Supergroup sediments without any break.

The dominant formation within the Dwyka Group consists of an enigmatic, fine-grained, bluish-black matrix studded with boulders of all sizes (some up to several tons in weight), pebbles and chips of a great variety of rock types that include sandstones, ancient schists, granites, gneisses, limestones, and diabases (altered lavas), (See Plate 7). This deposit reaches an impressive 600 to 750 metres in thickness in its southern outcrops. Exactly what its origin might be prompted an inordinate number of pages in the early geological literature – much of it entertaining, but completely wrong.

BAIN – THE FIRST MAN OF SCIENCE ON THE SPOT

Bain, the pioneer of South African geology, named this deposit 'claystone porphyry', which was perhaps an admission of defeat in establishing its origin. The term 'porphyry' is a descriptive one used by geologists for those igneous rocks (rocks crystallised from melts) in which relatively large, conspicuous crystals are held in a fine-grained or glassy base. Older usage of the term 'clay' was for 'an earthy deposit of extremely fine texture which is usually plastic when wet' (Rice, 1956: 77), and various names such as 'clay ironstones' and 'clay shale' were given to fine-grained rocks laid down by sedimentary processes.

The term was not one of local derivation coined by Bain, for it appears in the book from which he learned his science during his years of wandering the plains and mountains of the Cape Colony – an early edition of Lyell's *Elements of Geology*, in which 'claystone porphyry' is described as 'an earthy and compact stone, usually of a purplish colour . . . [containing] scattered crystals of feldspar and sometimes of quartz'. The name is therefore something of an oxymoron which could imply either an igneous or a sedimentary origin.

Hatch and Corstorphine (1905), authors of one of the first comprehensive books on the geology of South Africa, offered some interesting comments about Bain's interpretation of the origin of this

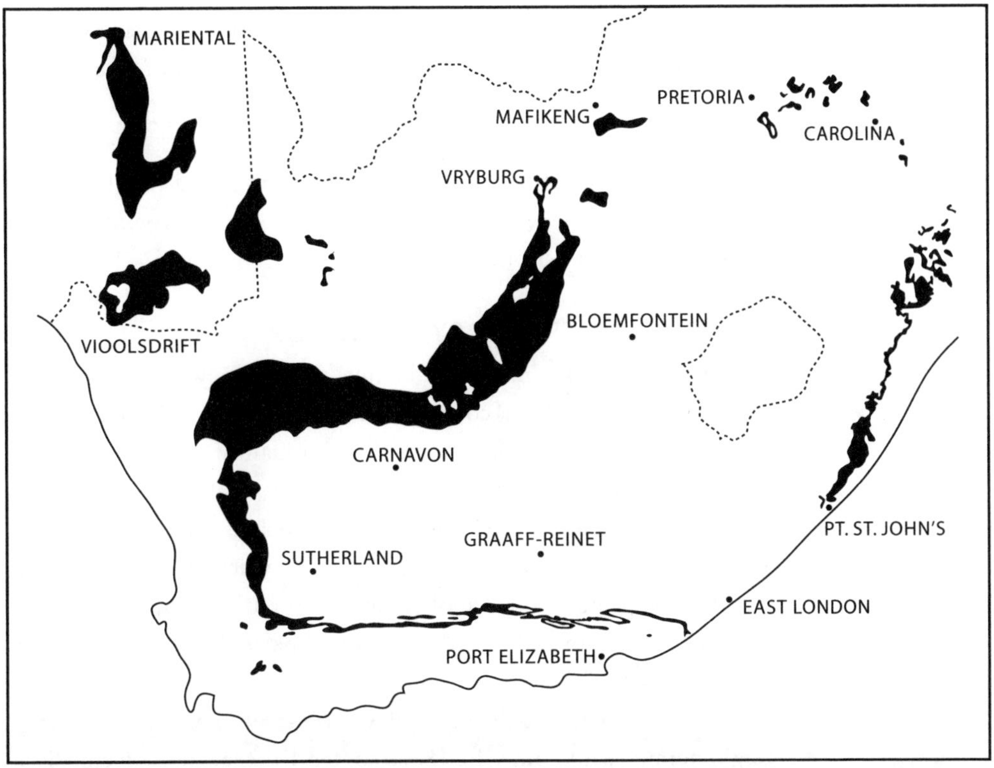

Figure 18: Distribution of the Dwyka Group within which occurs the Dwyka Tillite deposited under glacial conditions. The sinuous outcrop in the south, and its narrow width, are due to its being caught up in the steep folds of the Cape Fold Belt; further north, outcrops remain closer to the horizontal and spread over larger areas.

perplexing deposit. In eulogising his pioneering work they wrote: 'Practically the only error into which he [Bain] fell was in connection with a rock which puzzled many of his successors. His "claystone porphyry" is the much discussed and interesting Ecca or Dwyka Conglomerate, which he held to be "the production of an immense volcano" ' (1905: 3). If you were to follow this up in the *Transactions of the Geological Society of London* for 1856, you would find Bain's own rather lyrical account of this 'enormous igneous mass . . . whose northern limit perhaps extends thousands of miles into the interior' presented in the words:

> From the numerous imbedded pebbles of granite . . . one might be led to believe that the whole was an aqueous deposit . . . but this idea falls to the ground. I trust I may not be considered a visionary if I attribute the whole to the production of an immense volcano, which we may suppose to have existed somewhere near the junction of the Vaal and Orange Rivers . . . and

> thence deluged with fiery billows the Silurian (?) plain, and spread ruin and desolation over the carboniferous forest for tens of thousands of square miles.

Hatch and Corstorphine (1905: 3) added the remark that, 'while offering this explanation, however, he [Bain] admitted that he was in a dilemma, because he saw some indications that the rock had been produced by water action'.

A MAJOR CONTROVERSY FLARES UP

Bain was not alone in this dilemma. The review by Hatch and Corstorphine (1905: 3–16) covering the period 1856–1904 (summarised in the following paragraphs) makes this abundantly clear.

W.G. Atherstone, a medical man, fervent amateur geologist and friend of Bain, had in 1857 described the rock as 'vesicular' (a term used when gas bubbles are present, implying a volcanic origin), adding that 'it . . . appears as if the molten lava in bursting through the lower beds had broken up and mixed in its liquid paste the fragments of the different rocks it had passed through'.

Other pioneers such as Wyley, in 1859, named it a 'trap conglomerate', while Tate, in 1867, contended that 'it is decidedly a dolerite containing angular and rounded fragments of quartzite and granite' to which he assigned the name 'trap breccia'. These terms were equally inappropriate, as 'trap' is derived from the Scandinavian word *trappe*, meaning 'stairs' – referring to the stepped profiles of hills in Europe built of volcanic flows.

It was only in 1868 that P.C. Sutherland, the surveyor-general of Natal, recognised it as a rock of glacial origin after a visit to the Arctic. In 1870 Carl Griesbach, another early pioneer, concurred with this view, citing as supporting evidence the manner in which the deposits rested in places upon striated pavements (see Plate 8). Striations were well known in other parts of the world where ice sheets, armoured with embedded rock fragments, had advanced over and gouged scratches into the underlying rock surfaces.

Despite this, some famous names continued to hatch bizarre ideas (see Corstorphine, 1904: 145–81; Hatch & Corstorphine, 1905: 3–16). In 1871 Pinchin called it a 'metamorphic conglomerate'; Moulle in 1885 a 'melaphyre breccia'; and Professor A.H. Green (1888) of Oxford found it 'scarcely possible to draw any distinction between . . . [it] . . .

and some undoubted volcanic tuffs', but in the end decided it was a coarse shingle formed along a retreating coastline. In 1893 A.R. Sawyer stated that his microscopic study 'shows it to be undoubtedly a volcanic ash . . . the numerous volcanic pipes in the Karroo . . . are an indication of the source . . .' Sawyer was in good company, for even the eminent Sir Archibald Geikie in Britain concluded that it was 'a volcanic breccia' (see Corstorphine, 1899: 25–26).

In 1894, C. Wilson-Moore (see Draper, 1895) added to this clutter of ideas the whimsical view that it had been formed by submarine volcanism accompanied by a huge tidal wave that had swept from the south across the land. A few years later, David Draper (1897), a respected South African writer of the time, rather surprisingly endorsed this extraordinary interpretation.

In 1895 the influential Adolf Schenck gave the glacial theory a boost by stressing that the trapped boulders had been carried from sources as far afield as Bechuanaland (Botswana) and the Transvaal. Nevertheless, Draper (1895: 97) remained unconvinced, claiming that 'the glacial theory . . . cannot satisfactorily explain the circumstances under which this peculiar rock was formed'. The response by Molengraaff (1898), the state geologist of the Transvaal, made it quite clear what he thought of Draper's doubts by dismissing the latter's paper in a mere two lines of text, before going on to present field evidence that confirmed a glacial origin beyond any doubt.

Even the age of the deposits had been disputed. In 1887 Professor Cohen from Germany assigned them to the ice ages of the Pleistocene. Another visitor, Professor A. Prister, made the same mistake when he examined glacial phenomena around Pretoria and on the Witwatersrand. As Prister later apologised (1898a), '. . . the traces I enumerated in my paper were so fresh that I was misled into supposing that they belonged to a more recent [Quaternary] period'.

THE DISPUTE SETTLED AND CONSENSUS REACHED

By 1899 Dunn and Corstorphine had resolved many of the problems by recognising that there were genuine differences between what observers in the Transvaal and the Cape were seeing. In the south, this formation had undeniably been laid down under marine conditions by the melting of floating icebergs that had chaotically discharged their load of rock debris onto the sea floor. By contrast, in the north and in Natal the striated pavements proved that here it was a terrestrial sediment left behind by the melting of continental ice sheets studded with boulders and pebbles.

The 'Dwyka Conglomerate' had utterly fascinated Molengraaff (1904: 61–75) to the extent that he devoted fifteen pages of his ninety-page *Geology of the Transvaal* to it. He had not the slightest reservation about it being of glacial origin. If any further persuasion was still needed, R.B. Young and J.P. Johnson (1906) provided it in their descriptions and photographs of striations, polishing, and blocks of floor rocks plucked out beneath the ancient moraines at Riverton and Pniel, identical in form to those found in Europe.

TILLITES, CONTINENTAL DRIFT, ERRATICS AND DIAMONDS

In later years, this rock drew immense international interest when Alex du Toit used its widespread occurrence in Africa, South America, Australia and India to reconstruct the ancient supercontinent of Gondwanaland as it had been before its breakup into separate continental fragments (see Figure 12). He would have been neither impressed nor pleased to see a paper by C.G.S. Sandberg (1928) in the *Geological Magazine* (London). Here, Sandberg attempted to resurrect the argument once more by comparing the tillite with 'breccias of eruptive origin', and by pointing to striations beneath, and faceted and striated boulders within mudflows (lahars) and deposits laid down by fiery clouds (nuée ardentes) ejected from the volcanoes of Martinique and Java.

Sandberg's attempt to provoke a response from Alex du Toit by claiming that 'the modes of occurrence of the Dwyka deposit are irreconcilable with the glacial theory of its origin' was met with disdain. In a half-page reply in the succeeding issue of the same journal, Du Toit (1928) coldly dismissed Sandberg's paper with the words '... from the nature of his arguments I am reluctantly led to doubt whether much, if anything, is to be gained through attempting to dispose of his criticisms, point by point'.

It would be all too easy to deride these flounderings of the earlier geologists in this country, but a truer perspective emerges from recalling the turmoil that European geologists had been through early in the nineteenth century. Here, controversy had raged over the vast tracts of Pleistocene till (unsorted boulders in a fine-grained matrix) that covered parts of northern Europe and regions around the Alps

(Hallam, 1989: 87–102). Before the 1820s these had been referred to as 'diluvial deposits' – laid down by a great deluge, widely equated with the Biblical Flood. In time this had developed into the Drift Theory, attributing the unexplained presence of huge boulders of rock from distant sources (erratic blocks) to the melting of floating icebergs. The fact that this would have demanded sea levels as high as 2 700 metres in the Alpine regions was an echo of the misguided diluvial explanation. It had even been proposed that these erratic blocks had been 'shot violently through the air across the intervening Rhone Valley as a result of localised subsidence compressing air in deep caverns' (Hallam, 1989: 89), as if they were champagne corks!

Only in 1840 was a glacial theory of continental ice sheets put together by the Swiss geologist, Louis Agassiz. The most experienced geologists of Europe, enjoying all the advantages of field meetings and conferences, had taken decades to arrive at some degree of consensus in identifying glacial phenomena. Bain had had none of these advantages: he was a lone pioneer on the unmapped plains and mountain ranges of South Africa, publishing his work a mere decade after the controversy in Europe had come off the boil.

The outcrops continue to confuse even contemporary geologists. In the 1950s a visitor to the eastern Cape offered the scathing comment that the geologists there 'were asleep, apparently not having recognised the extensive outcrops of basaltic lava in the Ecca Pass' north of Grahamstown. To round off a catalogue of gaffes, in 2004 a sample, indisputably of tillite but labelled as 'kimberlite' (the volcanic host rock of diamond), was sent for evaluation to the geology department of Rhodes University by a reputable exploration group.

Perhaps, then, Bain's name, 'claystone porphyry', was not as ill-conceived as it appears, as it gave a hint of the controversy that was to rage for decades. The alternative terms that surfaced – ground moraine, dropstone conglomerate, volcanic ash, volcanic breccia, lahar, product of a submarine mud volcano, pyroclastic surge, trap breccia, porphyry, littoral beach deposit, basalt, product of a tidal wave, and even kimberlite – nicely emphasise the confusion of the past.

A resurgence of interest in the Dwyka Tillite has come from the suggestion that it could conceivably be the carrier of marine diamonds now occurring along and off the west coast of southern Africa, and of alluvial stones along the Orange (now Gariep) and Vaal Rivers.

J.M. Moore and A.E. Moore (2004) have raised the possibility that the continental ice sheets had acted like immense carpet sweepers that scraped diamonds off the ancient land surface where diamondiferous pipes had shed them. Subsequent erosion of the glacial deposits might then have released the diamonds that had been transported great distances by them. Here we might note that many years ago Fred Cornell (1920: 38), a well-known, old-time prospector, had advanced a similar idea for the origin of the west coast diamonds in his book, *The Glamour of Prospecting,* which recalls his experiences in the days when the ox-wagon and Shanks's pony were still the customary modes of transport in the rugged north-western regions of the Cape province.

12

The Cape and the Transvaal: A geological misalliance

'[Schenck, 1888] . . . unites the rocks of the Magaliesberg with those of the Witwatersrand, and correlates both with the Table Mountain Sandstone and Bokkeveld Beds, which in his opinion are equivalent facies. [His classification] has been more or less followed by subsequent writers.'

'It is a remarkable fact that . . . the siliceous chert layers [of the Transvaal Dolomite] are also devoid of fossil remains, notwithstanding the fact that these most probably originated from the accumulation of the siliceous skeletons of such organisms as sponges and diatoms.'

'Correlation of the pre-Karroo formations that occur in the various parts of South Africa involves more than ordinary difficulties . . . The sparse occurrence of fossils is not easy to understand.'

Hatch and Corstorphine (1905: 24, 161, 301)

Earth science was still on a shaky foundation in South Africa at the end of the nineteenth century, with practising geologists very thin on the ground, transport in the field dependent largely on ox-wagons or horses, and even reliable maps very incomplete. Aerial photographs did not exist, and the devastating Anglo-Boer War (of 1899–1902) had torn the country apart. The acting President of the Geological Society of South Africa, Molengraaff (1903) had reflected sadly, after the war, on 'three years of forced lethargy lying behind us', when 'a storm broke loose over South Africa, fierce beyond imagination, leaving the country . . . utterly devastated and ruined . . .'

In 1903, G.S. Corstorphine, the pioneering geologist with the Cape Geological Commission, was still lamenting the difficulties in obtaining the current literature. By the time that *The Geology of South Africa* (Hatch & Corstorphine, 1905) appeared, the Commission had devoted a bare six months to the northern part of the Cape Colony – the area around Prieska and Hopetown. When the Commission and the Geological Survey of the Transvaal were eventually merged into the Geological Survey of the Union of South Africa in 1910, there were only six geologists on the staff, with their running budget a meagre £10 000 per annum (Young, 1913).

Amongst the most urgent questions was this: are the rocks of the Transvaal, with their fabulous mineral deposits, a northward extension of those in the Cape, or do the two belong to 'different worlds'? The assessment by G.A.F. Molengraaff, the state geologist of the South African Republic, that '. . . the simplicity of the geological structure in its salient features still further assists the geologist' (1904: 1) was to prove to be an embarrassingly over-optimistic prediction.

The Cape Colony and the Transvaal had been separate countries, administered by different governments between whom relationships were, at best,strained. The bickering amongst a handful of geologists over the age of rocks would hardly have been regarded by state officials as of much, let alone national, significance. But, to understand the geological history of southern Africa, and then unearth its unique mineral deposits, the issues could not have been more fundamental.

Early perceptions were that a folded sequence of hard and compact sandstones, and marine shales that yielded fossils indicating a Middle-Palaeozoic age, existed along the southern margin of the Cape Colony. These rocks were grouped together as the Cape System, resting upon a floor of older slaty and schistose rocks (the Malmesbury Series) invaded by the Cape Granites. North of this stretched the widespread Karoo System, recognised as an overlying blanket that reached the southern Transvaal, where older and strongly folded rocks emerged once more from beneath it.

The area covered by this 'blanket' is enormous (more than 620 000 km^2; see Figure 19) and its thickness approaches a massive 11 kilometres beneath a capping of lavas in the Drakensberg Mountains. At the time, it seemed natural enough to couple together the older formations exposed respectively in the north and south, as Adolf Schenck, Professor at the Halle University in Germany, had done in 1888. The rocks of the Cape System in the south were paired with those of the Witwatersrand (the sequence from which most of South Africa's gold was being mined), together with those overlying it, known today as the Transvaal Supergroup (see the table on page 111).

The dilemma that this led to emerges in the extracts quoted at the start of this chapter, taken from *The Geology of South Africa* by Hatch and Corstorphine (1905).

AN UNEASY MATCH

At the time, most of the leaders in South African geology accepted this match. In the early 1890s, the respected David Draper had regarded the resistant strata building the prominent Magaliesberg Mountains, west of Pretoria, as the northerly equivalent of the Zuurberg and Witteberg ranges in the Cape province, and the underlying Witwatersrand rocks as the equivalent of the 'Table Mountain Series', the bastion that rears up above the city of Cape Town, and stretches northwards and eastwards from there (see Plate 9).

To justify this supposed match, Draper needed to find, in the Transvaal, the equivalent of a thick succession of richly fossiliferous rocks (known then as the Bokkeveld Beds) that lie within the 'Cape System'. Draper (1897: 19–30) asserted that this was the thick carbonate sequence (the 'Dolomite Series') in the Transvaal, in his view laid down when 'the oceans teemed with life . . . and banks [were] built up by myriads of polyps . . . to be subsequently changed by chemical action, accelerated by the tropical heat of the waters, into dolomite'.

This was a profound misconception – a flight of imagination. Corals do not in fact appear in the geological record until early in the Palaeozoic, some 2 billion years later, and the dolomites of the

Transvaal (up to 3 km thick, in which dangerous sinkholes develop) are in reality not fossiliferous, apart from the remains of primitive colonies of algae.

Draper was not alone in repeating Schenck's mistakes. Molengraaff wrote that 'the only point on which . . . opinions are fairly unanimous, is the recognition of three principal divisions in accordance with Schenck' (1897a), and on this basis he compressed the geology of South Africa into three great divisions: the South African Primary Formation, the Cape Formation, and the Karoo Formation. Convinced that 'the lowest division of the Cape formations consists at Cape Town – and so far as our knowledge now extends – over all South Africa', he offered the following model for everything that lay beneath the Karoo Formation:

(IN THE SOUTH AFRICAN REPUBLIC)		(IN THE CAPE COLONY)
CAPE SYSTEM	Magaliesberg or Pretoria Series	Zuurberg or Witteberg Series
	Dolomitic Limestone	Bokkeveld Beds
	Black Reef Series	Table Mountain Sandstone
SOUTH AFRICAN PRIMARY SYSTEM	Witwatersrand Series	
	Intrusive Granites	Intrusive Cape Granites
	Barberton Formation	Malmesbury Beds

By 1904, Molengraaff had come to doubt this, his own model. In prophetic words, he declared: 'The age of the South African Primary System in the Transvaal is unknown . . . most probably the [Transvaal System] is older than the Cape System . . . and our Rand rocks older again' (1904: 18).

RED HERRINGS IN THE PATH OF THE EARLY PIONEERS

We must ask what had led the early pioneers to be so wide of the mark. Their work had implied that the hard sandstones building Table Mountain (see Plate 9), in reality less than 450 million years old, were the southerly equivalent of the Witwatersrand and overlying sequences, now known to range in age from around 3 000 to 2 100 million years.

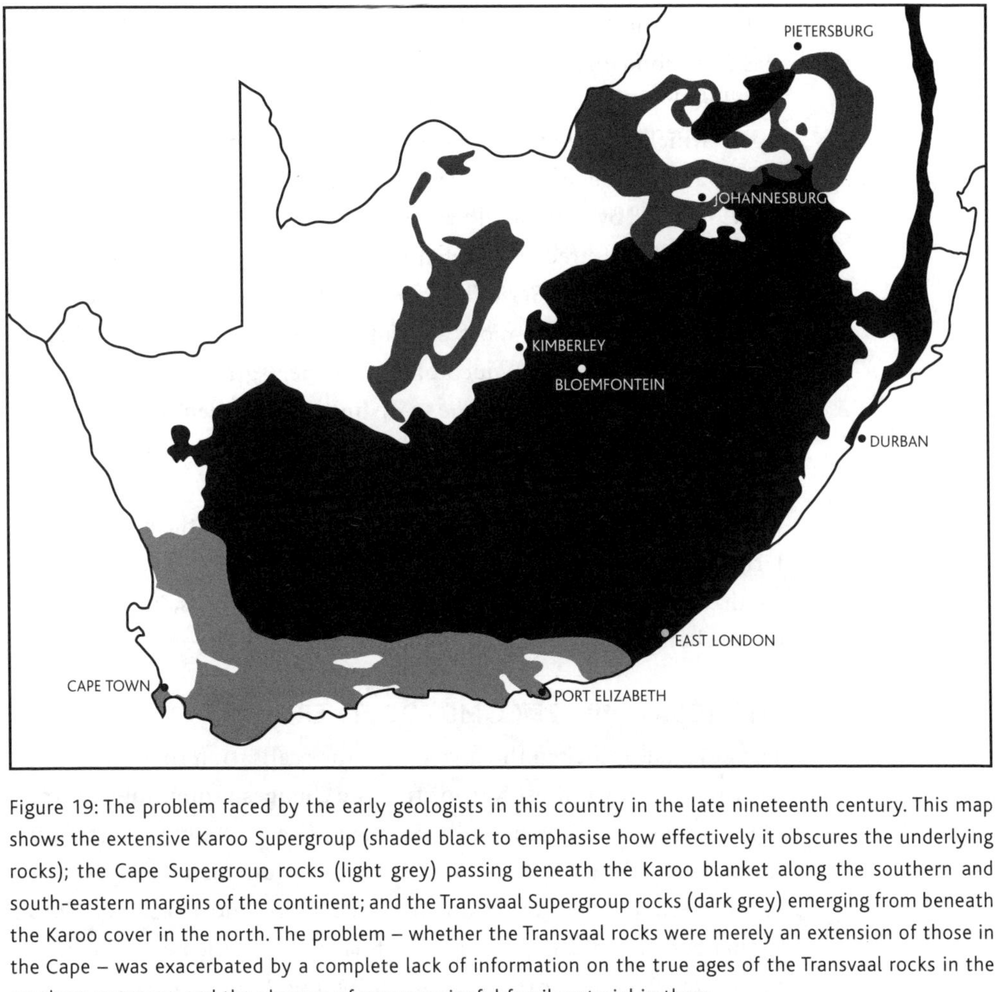

Figure 19: The problem faced by the early geologists in this country in the late nineteenth century. This map shows the extensive Karoo Supergroup (shaded black to emphasise how effectively it obscures the underlying rocks); the Cape Supergroup rocks (light grey) passing beneath the Karoo blanket along the southern and south-eastern margins of the continent; and the Transvaal Supergroup rocks (dark grey) emerging from beneath the Karoo cover in the north. The problem – whether the Transvaal rocks were merely an extension of those in the Cape – was exacerbated by a complete lack of information on the true ages of the Transvaal rocks in the northern outcrops, and the absence of any meaningful fossil material in them.

There were several reasons for this mismatch. The absence of marine Palaeozoic fossils in the dolomites of the Transvaal was an enigma in the eyes of Hatch and Corstorphine (1905: 302), although they had clung to hopes that 'fossils may still be there, which further search one day will bring to light'. With the benefit of hindsight, we know that this optimism was a vain hope – the fossil sponges they sought were not due to appear in the rock record before the passage of more than another 2 billion years. Nor were matters helped by the red herring of unfounded reports in 1885 about fossil graptolites (extinct marine invertebrate animals) and corals having been discovered in Transvaal rocks. Molengraaff (1904) muddied the waters even further with his claim to have found fossil crinoid stems – a resounding faux pas from someone of his standing.

Another red herring is attributable to coincidence. The foundations upon which the widely separated exposures in the Cape and the Transvaal happened to rest were, in both cases, old slaty and schistose rocks invaded by granites. To the early pioneers, the geological settings would have appeared to be identical. In the Cape, these foundations are the 'Malmesbury Series' invaded by the Cape Granites; but in the Transvaal they are our most ancient Archaean rocks (the Barberton Supergroup) invaded by much older Basement granites. There was no technique available at that time to distinguish between them.

In the succeeding decades the older ideas were at last completely overturned and, by 1929, with the publication of *The Geology of the Union of South Africa* by Rogers et al, it was clear that the history of the terrane north of the Vaal River was in reality very different to that in the south. Geologically, they were, indeed, different worlds, but the details of the change were deeply hidden under the thick Karoo blanket.

THE JIGSAW PIECES COME TOGETHER

The key that unlocked the puzzle was the realisation that the surface on which the lowermost Karoo strata lie changes profoundly between the south and the north, and that this is one of the most significant surfaces in all of southern Africa. In the south there was no great hiatus between the laying down of the Cape and overlying Karoo Sequences (technically, they are conformable). In the north, however, a profound gap in time (an unconformity) opens up between the Karoo strata and whatever older rocks they happened to be draped over. This unconformity may in places represent more than 3 billion years of erosion and unrecorded history that was finally brought to an end by the massive Dwyka glaciation that overwhelmed Gondwanaland in the Carboniferous Period.

Your reaction to all of this turmoil could well be, 'So what? Was this of any greater consequence than the mistake the early trekkers ('travellers') had made in claiming that they had discovered the source of the Nile in the northern Transvaal?' Squeezing the history of southern Africa into the short span of time since the Silurian or Devonian was, however, a fundamental error that would have held back the interpretation of its history for as long as the idea survived.

It would have been futile to have scoured the rugged mountains of the southern Cape for the asbestos, iron and manganese ores that lie further north. How much effort would have been wasted searching in the Cape for the riches of the Witwatersrand goldfields, spurred on by the trivial occurrences of gold in the southern Karoo? Gold had, for example, been reported at Prince Albert.* Would prospectors have looked in the Cape for another Bushveld Complex, with vast reserves of platinum, vanadium and chromium? Would palaeontologists have persisted in a dogged but vain search in the dolomites of the Transvaal for marine fossils like those in the Cape System? The geological history of southern Africa would have been an incomprehensible tangle. Geologists would have been as bemused as historians might be by claims that the pyramids of Egypt and the Great Zimbabwe Ruins were of the same age.

* A report by Atherstone and Bain on the prospects of mining gold in the Prince Albert area, dated July 1871, was commissioned by The Honourable Richard Southey, Colonial Secretary of the Cape of Good Hope, following the reputed discovery of a gold nugget weighing 2½ ounces. A nugget did, in fact, exist, and was examined by the authors of the report, but it is left to the reader to interpret what they implied by 'we came to the conclusion that it had not been derived from any rocks in the neighbourhood, nor could it have been transported there by natural agencies'.

At a general meeting of shareholders of the Prince Albert Goldfields Prospecting and Proprietary Syndicate, the Secretary (Thiel, 1906) summarised a 'not at all unfavourable' report by the consulting geologist, Dr Sandberg, which 'alludes to this part as the most interesting in the Colony . . . [in which] we have a belt of gold-bearing formation, consisting of a sandstone reef from four to 40 feet thick, intersected by thin veins of quartz'. The Chairman then told the meeting that 'our manager has discovered that the gold belt is much wider than anticipated, fine gold being found a considerable distance away, north and south of our operations . . .' The gold found by the syndicate had to date amounted to a mere 14 ounces.

13
E.H.L. Schwarz's Kalahari Scheme

'. . . I have been able to put forward a plan for conserving the waters of the Cunene, Okavango and Chobe Rivers, and impounding them in a great natural reservoir, the Makarikari . . . a vast lake would exist to supply South Africa with moisture, and stop the desiccation that is ruining the land.'

'All down the course from the Ngami to the Orange River, below the falls, settlements will arise and agriculture, of the same nature and on the same scale as that in Egypt, will spring up.'

'. . . the arguments that I have brought to bear on the subject and the results to be expected from the Kalahari Scheme are not in any way theoretical, or based on unproved facts.'

E.H.L. Schwarz (1920: 102, v, iv)

'I am afraid that the conquering of the Kalahari will have to be effected not by such a grandiose scheme involving the restoration of the lakes, but by slower and less spectacular means.'

Alex du Toit (1927b: 100)

'I happened to fall into conversation with General Smuts [the prime minister at the time] about the Okavango swamps and the Schwarz theory, and I told him there was nothing in it as far as increasing our rainfall was concerned . . .'

Deneys Reitz in No Outspan (1943)

SCHWARZ'S KALAHARI SCHEME

Schwarz's contention was that the waters inside the barrier-ranges of the great escarpment 'once flowed towards Central South Africa, and made the central depression a land of running rivers and of great fertility' before headward erosion by the rivers below the escarpment breached the divides and 'hurried the waters uselessly to the sea' (1920: 141–42). According to his reconstruction, the Loangwa River that rises in the highlands west of Lake Malawi (Nyasa) had flowed southwards to join the present course of the Zambezi River, where at that time it turned ***westwards*** to flow towards Wankie (the reverse of its present direction of flow), before draining into a great lake in the Makarikari (Makgadikgadi) Depression of Botswana (or Bechuanaland Protectorate, as it was then known). From here, he held, the flow continued through Anderson's Vlei further south, to join with the Molopo River and eventually reach the sea along the course of the Orange (now Gariep) River.

Schwarz was convinced that the climate of southern Africa could be radically changed and restored to what it had been for centuries prior to 1820 by once again flooding vast parts of the interior with water to create inland lakes, through damming and diverting the waters of major rivers such as the Cunene in Angola, and the Chobe (Linyanti or Kwando) and Okavango in Botswana.

The waters of the Cunene were to be diverted into the great depression of the Etosha Pan in the west by building a weir above the smaller cataracts upstream from the greater Ruakana cataracts. The Okavango was to be diverted into the dried-up Lake Ngami, at one time a sheet of water some 300 by 100 miles (480 x 160 km) in extent before it was drained, according to Schwarz, by headward erosion and river-capture by the Zambezi River system. Likewise, a weir on the Chobe River was to divert the water into the Makarikari Depression.

The scheme proposed by E.H.L. Schwarz in 1920 was, in a nutshell, to divert some of the major rivers of southern Africa into natural depressions in Botswana, and create lakes there, from which evaporation was to increase the humidity of the atmosphere, and thereby increase the rainfall. As Alex du Toit (1927b) put it: 'he formulated his hypothesis of the former existence of a great lake in the northern Kalahari . . . "restore those lakes" was his slogan'.

SCHWARZ'S WARNING – AND HIS SOLUTION

Most people would believe that Schwarz was quite justified in his analysis that indicated a progressive drying out of southern Africa during the preceding 200 years. He had drawn this conclusion from a variety of observations: the study of rainfall and rates of evaporation; the comparison of descriptions of the land written by the earliest travellers and explorers in southern Africa with what was before his eyes at the beginning of the twentieth century; the silting up of rivers; the dry river beds in which water rarely flowed; increasing levels of salts accumulating in the soils; the progressive expansion of the semi-arid Karoo ecosystem, and so on. (The same conclusion might perhaps be drawn today by noting the hundreds of farms across the length and breadth of South Africa with names terminating in '-vlei' [meaning 'wetland'] or '-fontein' [meaning 'fountain'] – Rietfontein, Rietvlei, Sterkfontein, Tweefontein, Driefontein, Modderfontein, Vlakfontein, Palmietfontein, etc. – many of which are today arid, unproductive and even abandoned tracts swept by hot, dry winds.)

Commenting on the £900 million that had been yielded by exploiting the mineral deposits

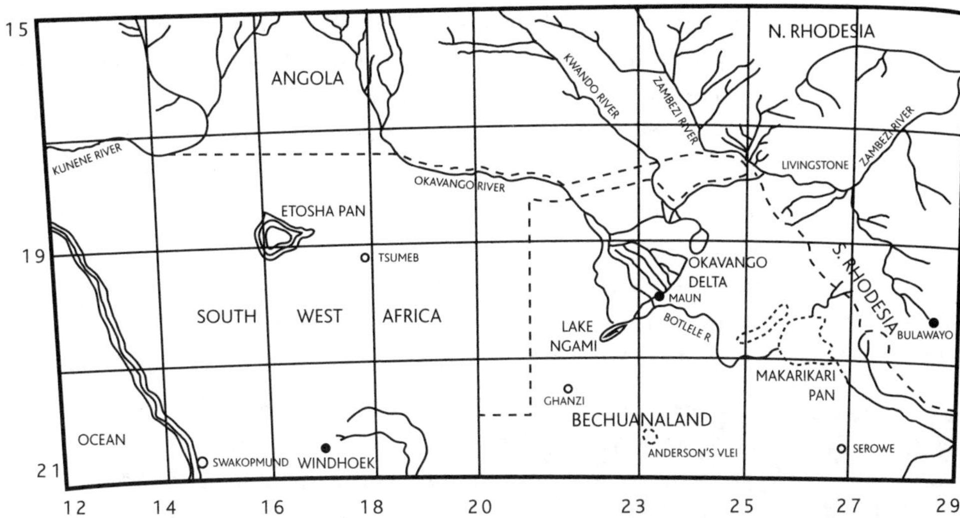

Figure 20: Map of the region that E.H.L. Schwarz sought to transform by the diversion of major rivers into natural depressions (Etosha Pan, Lake Ngami and the Makarikari Depression) in the belief that this would increase moisture in the air, and hence precipitation. The map is a simplified version of the original accompanying Du Toit's report to the Department of Irrigation in 1925, and preserves the place names and their spelling that he used at the time.

in South Africa through the preceding fifty years, Schwarz wrote, 'in another fifty years, what remains will be exhausted. What then? The enormous debt of the country will fall upon the shoulders of the impoverished sons of the soil . . .' (1923: 221).

Schwarz calculated that the flooding of the Etosha Pan and Makarikari Depression (see Plate 10) would inundate an area of 52 000 square kilometres, which, given an average depth of 3 metres, would hold 155 billion cubic metres of water. In this way, the ratio of precipitation to evaporation in southern Africa (which he claimed to be 1:3) would be improved and initiate a continuous cycle of increased rainfall and replenishment of moisture retained in the soils.

At the time this seemed to be a great idea, especially if considerations such as the riparian (riverbank) rights of countries downstream, or the expulsion by the rising waters of whole communities of indigenous peoples, not even living within the borders of the Union of South Africa, were simply ignored. However, even during the earliest stages following the publication of Schwarz's scheme, it was judged by many people to be a doubtful proposition, and a report by a government commission in February 1921, led to any further consideration of the scheme being abandoned. This was how matters stood until July 1924, when the offer of £1 000 from an anonymous donor rekindled sufficient interest for a team of surveyors and engineers, under the leadership of Alex du Toit, to be appointed by the Department of Irrigation to have another look at the scheme.

DU TOIT'S DEMOLITION OF THE SCHEME

The investigation (Du Toit, 1925) was as thorough as a limited budget would allow. Accurate spirit levelling was extended over 1 700 kilometres, the first accurate determinations of longitude were made, and the region was surveyed by photographs taken from a small aircraft. While acknowledging the Kalahari as 'the largest continuous area of sand known anywhere in the world', Du Toit (1927b: 90) maintained that its area was actually shrinking, and contended that 'only through an improvement in the rainfall could that desert, once like that of Libya or the Namib, have been retrieved' to allow the present cover of grass, trees and shrubs that now anchors the sands to have become established.

Du Toit's analysis suggested that the Okavango River would in the past have maintained its general south-easterly course over the north-eastern corner of the Kalahari, continued to flow along the Botletle (Botetei) River to an immense lake formed by the Makarikari, and then discharged via the Macloutsie River into the Limpopo River, had there not been a significant disturbance of the Earth's crust in that region. At that time, the Okavango Delta would not have existed. Du Toit attributed the subsequent shift in the

Figure 21a: The natural depressions that Schwarz sought to flood by the diversion of major rivers in the interior of southern Africa, and thereby increase the humidity and rainfall of the Kalahari region.
Figures 21b: The Makarikari Depression (or Makgadikgadi Pan) under investigation as a source of natural salts.
(Photographs by courtesy of Professor R.E. Jacob.)

course of the Okavango River to warping of the land surface, rather than to climatic factors as Schwarz had done. Du Toit identified upwarping along well-defined axes, linked with the development of the Great Rift Valleys of East Africa, as the process that had isolated the Makarikari Depression from the Limpopo catchment area. Sagging of the crust along a northeast-trending depression some 160 kilometres wide had created a lower-lying trough into which the Okavango then flowed, and in which the delta then began to grow, depriving the Makarikari Depression of water. Thus, he argued, the evaporation in the past would have been much as it is today, and no improvement in the rainfall could be anticipated.

In addition, in his 1925 report Du Toit identified immense engineering problems that would be encountered. Where Schwarz had determined relative heights of the land surface as best he could by measuring the boiling point of water, Du Toit claimed that 'differences in level that were supposed to be small prove from actual survey to be embarrassingly large'.

Perhaps Du Toit's most telling point was his conclusion that the wastage of water by evaporation would not be compensated by increased humidity in the atmosphere, as Schwarz believed. Du Toit pointed out that actual measurements had shown that along the borders of immense swamps the humidity of the air was still low, and that rainfall adjacent to such 'evaporators' was no higher than at points further away to east or west. As he put it: '. . . the whole of the flow of the Okavango and much of the Linyanti [Chobe] is being evaporated in the north-eastern Kalahari, without any obvious influence on the precipitation within Ngamiland'.

Thus, between 1925 and 1927, the doyen of South African geology demolished virtually every basic tenet of Schwarz's scheme. Had Schwarz been more analytical, he might have paid greater attention to the fact that the rainfall is not enhanced by evaporation into the atmosphere around the Great Lakes of the USA and Canada, or around the vast Caspian Sea. Schwarz had contended, rather, that conditions in Africa are different.

What Schwarz had intended should be his priceless gift to southern Africa – that of increased rainfall – ended up instead as further testament to the uncanny perspicacity of Du Toit. On a budget that was trivial by today's standards, he had unravelled the geological history of this region and drafted it into a major scientific paper in 1933 – an accomplishment viewed by Moore (1999) as 'a remarkable paper, with important implications for understanding the geomorphological evolution of South Africa'.

Du Toit's analysis had anticipated, in essence, the results of modern surveys using satellite technology, aerial photography, geophysics and the drilling of boreholes (Gumbricht et al., 2001). The Okavango Delta, nearly 200 kilometres in width, is now recognised as occupying a relatively low-lying trough between more elevated shoulders to the north-west and south-east, while other, older fossil deltas are located several hundreds of kilometres further north-east and south-west of it. The uplifted terrain responsible for the disruption of the drainage systems is now termed the Ovamboland–Kalahari–Zimbabwe axis by Moore (1999), sweeping like a great wrinkle across the face of southern Africa.

The Okavango Delta itself is a relatively young feature, regarded as Late Pliocene or Early Pleistocene, with seismic activity suggesting that the depressed area may be bounded by faults along which fracturing of the crust is still taking place. There is also evidence that, at one time, the sheets of water that Schwarz had envisioned had been even more extensive than he imagined, with Lake Ngami and the Makgadikgadi and Mababe Depressions once an immense, continuous sheet of water in central southern Africa.

14
THE BUSHVELD COMPLEX: COMET IMPACT, MANTLE PLUME, OR LAVA LAKE?

'I have given this name [Plutonic Series of the Bushveld] to a group of intrusive and eruptive rocks . . . the intrusion of which took place at a period subsequent to the formation of the Pretoria Series . . .'

G.A.F. Molengraaff (1901) as quoted by A.L. Hall (1932: 26)

'. . . an essentially sheet-like intrusion of basic rocks . . . emplaced on a floor of sediments and roofed in part by sediments . . . and in part by [volcanics].'

A.L. Hall (1932: 151)

'. . . it is difficult to come to any other conclusion than that the basic rocks originated by the transformations of a series of sediments . . . by the introduction of atoms and ions . . . in a rising front.'

S. van Biljon (1958)

'I favour an extrusive origin . . . an unpublished survey . . . gives me a strong impression that tongues of Bushveld mafic rocks projecting into the Pretoria sediments are not intrusions but infillings of "drowned valley" and that the contact between the two formations is similar to that of a drowned coastline, where the mafic rocks acted as the sea and the Transvaal System the sea floor and coastline.'

C.A. Cousins (1959)

'The Bushveld Igneous Complex is interpreted here to have formed in three overlapping hypervelocity-impact craters . . .'

Warren Hamilton (1970)

'This model proposes the rising of a plume of partially molten material from the mantle, the world-wide layer that underlies the crust. The heat of this was adequate to melt the overlying crust. Since mantle rocks and crustal rocks have very different compositions, the magmas they yielded became the respective parents of the layered sequence and of the low-density granites that overlie it.'

C.J. Hatton and J.K. Schweitzer (1995)

The Bushveld Complex is without question the 'jewel in the crown' of South Africa's geological phenomena. It is the greatest intrusion of its type in the world, and also a treasure house holding unmatched reserves of exploitable ores. It is a mass of rock that crystallised from a huge injection of molten liquid, at initial temperatures above 1 250 °C, into the crustal rocks underlying the Free State, Mpumalanga, Northern or Limpopo, and North West provinces of South Africa.

As far back as 1932, A.L. Hall, the Assistant Director of the South African Geological Survey, published a compelling 560-page memoir on the Complex. The following visionary words appear in his Introduction, which he then rounded off with a quotation that would prepare his readers for the puzzle he was about to offer them:

The Bushveld Igneous Complex is a vast treasure ground which for many years will provide the mineralogist, the petrographer and the geologist with a rich field to dig over. In submitting to the indulgence of the reader the following attempt as a first approximation to the truth, the writer would recall the words of a great thinker:

'Nothing would be done at all if a man waited till he could do it so well that no one could find fault with it . . .' (Hall, 1932: 19).

The most conspicuous feature of the Bushveld Complex is the way it crystallised, not as a simple, homogeneous mass, but as a stacked sequence of alternating darker and lighter layers of crystalline material, individually millimetres to tens of metres thick, building a pile ultimately reaching 7 to 9 kilometres in all. This is, in turn, overlain by a further 3 to 4 kilometres of granite. Amongst these layers are those bearing the greatest platinum reserves in the world, enormous tonnages of titanium- and vanadium-bearing iron ores, and layers of chromitite that hold some 50 to 60 per cent of the world's known sources of chromium metal.

The Bushveld Complex is divisible into five limbs or lobes (see Figure 22). The distance between the extremities of the limbs along a line from west to east reaches around 500 kilometres, and its width from Villa Nora in the north to the Bethal area in the south a little under 400 kilometres. Sheer size has been a daunting factor in unravelling its structure and the history of its emplacement a little more than 2 060 million years ago, and in accounting for the diversity of the rock layers and mineral deposits within it.

Inevitably, as its interior was penetrated by the development of its mines, hypotheses about its origin materialised like birds following a plough turning over new soil. Here was something to stretch the imagination to its limits – the amount of heat energy needed to produce an estimated

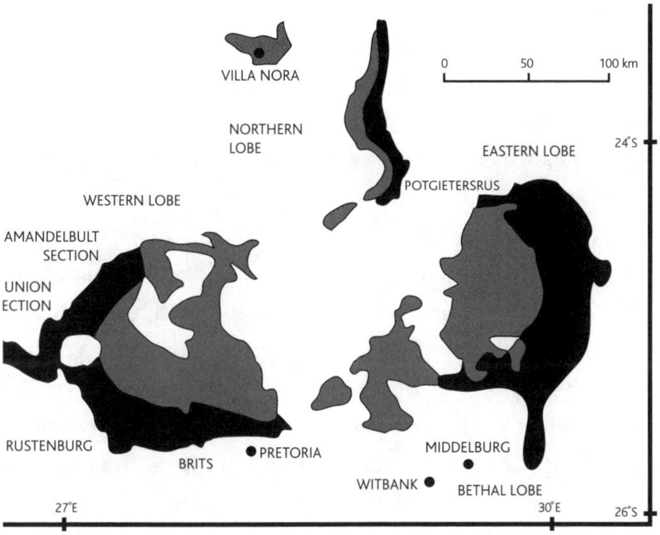

Figure 22: The huge Bushveld Complex, the largest igneous intrusion of its kind in the world. Shown in black are the layered rocks – largely a pile of dark, heavy rocks reaching a total of 7 to 9 kilometres in thickness. The overlying Bushveld Granites, 2 to 4 kilometres in thickness, are shown in grey. The Western and Eastern Lobes dip in towards the centre, and the Northern Lobe dips to the west, so only the edges of the intrusion are exposed at the surface. The Bethal Lobe in the south-east is wholly covered.

1 million cubic kilometres of rock and inject it into the upper crust over a period of some 75 000 years (Cawthorn & Walraven, 1998). The heat energy draining out of this into the surrounding envelope of rocks through the succeeding 200 000 years, while the liquid solidified, was enough to raise their temperatures until hundreds of metres of host strata were themselves recrystallised. All of this was done under sufficient pressure to lift a roof, considerably more than 5 kilometres thick, by some 7 to 12 kilometres. Not surprisingly then, many different reconstructions of this 2-billion-year-old event have found their way into print through the years.

AN ABUNDANT HARVEST OF IDEAS

The literature in this field is daunting in its sheer volume, with the extracts quoted at the start of this chapter representing only a small sample and expressing some of the better-known hypotheses, including: a huge intrusion giving way to lavas above; a huge intrusion emplaced beneath older lavas and sediments; a sequence of originally sedimentary rocks; a succession of lavas extruded at the surface; the result of impact with a comet; and the layered succession produced by magma generated in the mantle, while the crust melted to yield the overlying granites.

Amongst the early pioneers, Hall and Molengraaff had no doubts about the general features and the intrusive nature of this enormous mass of rock. From a geological map published by Molengraaff in 1901 (see Molengraaff, 1904) it is clear, however, that appreciation of the full extent of the layered rocks, and their origin, had scarcely begun at that date.

Molengraaff had interpreted the Complex as a mass of intrusive 'Red Granite' whose composition changed with depth, but in which 'the distribution of the magma was such that towards the periphery it became more and more basic [that is, darker and heavier], while towards the central and higher part this magma was acid and granitic [that is, lighter in colour and density]'. The wide expanses of layered dark rocks that we now know to be exposed over many hundreds of kilometres were represented on his map as no more than small and scattered patches that he believed had segregated from the granitic liquid. We could hardly expect his representation to have been much different at a time when 'of this enormous extent scarcely twenty square miles [(50 km^2) had] been examined' within what he estimated as a total area of 23 000 square miles (60 000 km^2) of outcrop.

Then came a period during which the seemingly obvious interpretation of the Complex as a very large igneous intrusion had to weather onslaughts provoked by maverick ideas which enjoyed some credibility before their day passed.

The origin proposed in 1949 by Van Biljon in a lengthy paper of 175 pages was in a class of its own. The finer details of his proposition will be best appreciated by practising geologists, but even those who are not will recognise the iconoclastic nature of his contribution. It was Van Biljon's contention that inside the girdle of (Transvaal Supergroup) sediments that now host the Complex, sedimentary rocks had been transformed into rocks that only happened to look like, and have the chemical composition and textures, of igneous rocks. The process he envisaged was metasomatism, permeation of the sediments by hot, reactive fluids and vapours that reconstituted them and changed both their composition and appearance.[5]

The reaction of the geological community to this proposition was one of shock, both in South Africa and overseas. The ideas flew in the face of all the theory evolved by petrologists through considerably more than a century. On receiving a pre-print of the paper, Professor R.A. Daly (1949: 177) of Harvard University asked '. . . should not the author realize that geologists the world over are going to be very critical of this lengthy paper?' His one-page commentary was a clear warning that Van Biljon would be doing himself a disservice by publishing it. F.C. Truter, who rose to become the Director of the Geological Survey of South Africa six years later, commented (1949: 178) that 'the transformations postulated by the author are truly astounding'. Professor B.V. Lombaard (1949: 181) of Pretoria University, a leading authority, was equally critical, referring repeatedly to 'arbitrary conclusions'.

Surprisingly, the doyen of South African geology, Alex du Toit (1954: 194), was less critical in his appraisal, commenting that 'while revolutionary, this hypothesis seems competent to meet not a few of the serious difficulties under current theories . . .' During the period 1970–90 a team drawn from various universities sponsored by the Council for Scientific and Industrial Research (CSIR), acting in tandem with the mining fraternity, researched various aspects of Bushveld geology. Not a single supporter of Van Biljon's self-confessed 'heresy' could be found.

C.A. Cousins's view (1959: 198) that the Complex was a sea of lavas extruded at the surface was heresy of a different sort. The extract presenting his interpretation, reproduced at the start of this chapter, did not actually appear in the body of his paper on the Bushveld Complex – he was pushed into publishing it in reply to sharp questioning from F.C. Truter, in the discussion that followed his presentation. Because the magmas of lavas are in general rapidly chilled when extruded at the surface, and crystallise at pressures very close to atmospheric pressure, they show in detail features rather different to those that crystallise more slowly at depth. Citing textural and other evidence to show that the rocks 'have consolidated at considerable depth', Truter would have received almost universal support for his assertion that 'nothing can be further from the truth than that the mafic rocks of the complex are extrusive'.

Ten years after Cousins's presentation, and twenty after Van Biljon's, yet another radically different hypothesis was launched by Warren Hamilton (1970: 367) of the US Geological Survey in Denver, Colorado, at a time when interest in structures produced by the impact of large meteorites on the surface of the Earth was growing in the USA. The Sudbury Intrusive of Ontario, Canada, had recently been reinterpreted as an example. It was asserted that great volumes of molten magma had been produced by the energy released by hyper-velocity impact, and that this magma had flowed into and filled a crater thus formed.

A similar model for the origin of the Bushveld Complex was then proposed by Hamilton, but the impacting body here was suggested to be a comet rather than a massive meteorite, and the different lobes of the complex the result of collision not with a single extraterrestrial body, but with three separate bodies. Hamilton claimed that the impacts were so violent that folding of rocks up to 200 kilometres away from the centres of impact could be attributed to the event.

The reaction of South African geologists to Hamilton's hypothesis was lukewarm. Surprisingly, where Cousins (1970: 374) might have greeted certain aspects of it with acclaim, in so far as they tended to support his notion of a volcanic signature to the rocks of the Complex, he totally rejected the idea. Cousins identified the folding of the surrounding rocks that Hamilton had attributed to impact as having already been in existence long before the alleged comets had struck. The siting of the different limbs of the complex dead-centre in the sedimentary basins was also seen as too much of a coincidence for them to have been random strikes by meteorites, since 'meteors to have hit three bullseyes would have had to be *aimed*'. Furthermore, G. von Gruenewaldt of Pretoria University (1970: 376) asserted that the lavas (known as the Rooiberg felsites) overlying the Complex were older than the underlying rocks that Hamilton had attributed to melting by impact.

Close on 30 years later, Paul Buchanan and Uwe Reimold (1998: 149), two international experts in the field of impact-related phenomena, poured more cold water on Hamilton's interpretation in the words: 'Extensive field work throughout the Rooiberg Group . . . has failed to identify any material that could be interpreted as impact-related.' After reviewing the field, geochemical, petrographic and textural evidence they concluded that: 'These data . . . are inconsistent with formation of the Bushveld Complex by impact.'

These days, even if there are certain aspects with which they might disagree, the majority of geologists would incline towards the reconstruction by C.J. Hatton (of the Anglo-American Corporation) and J.K. Schweitzer (of the CSIR), published in 1995. As illustrated in Figure 23, a mass of partially molten material (termed a mantle diapir) is believed to have risen slowly towards the surface through the mantle underlying the crust. On intersecting the crust, widespread melting of its rocks occurred, to produce at depth a pool of low-density, SiO_2-rich (granitic) liquid floating upon a reservoir of heavier liquid (more enriched in iron and magnesium oxides) that had separated from the mantle diapir itself. Injection of these liquids into a higher level in the crust then gave rise to the Complex, with the heavier liquid crystallising as the sequence of layered rocks, and the lighter liquid as the granites that today overlie them.

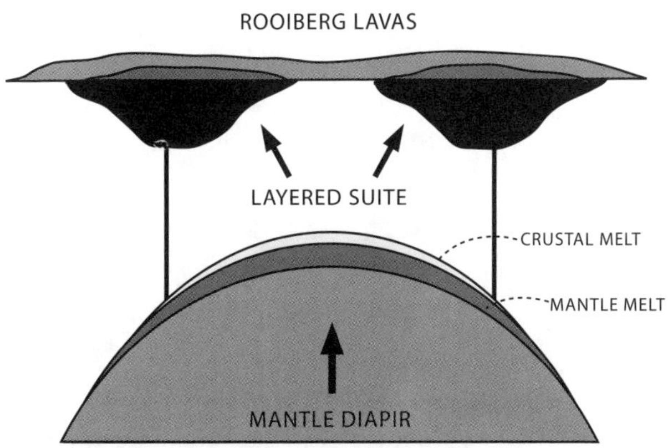

Figure 23: The emplacement model of Hatton and Schweitzer (1995). A rising mass of hot, partially molten mantle rock (a mantle diapir) is considered to have yielded a pool of liquid that was parental to the layered suite of the Bushveld Complex (shown black) while the melting of crustal rocks yielded an overlying reservoir of lighter magma that was parental to the Bushveld Granites.

This brief sketch of the Bushveld Complex is a preamble to Chapter 38 where the focus is on the origin of its unique deposits of metals. Even A.L. Hall himself, having predicted in 1932 that here was 'a vast treasure ground . . . with a rich field to dig over', could not have anticipated the developments that were to take place through the latter part of the twentieth century.

15
THE PRETORIA (TSWAING) SALTPAN: VOLCANO OR IMPACT CRATER?

'The whole feature bears so striking a resemblance to a crater of the Maar type that its volcanic origin was suspected even by laymen many years before it was actually established.'

P.A. Wagner (1922: 9)

'All of these findings . . . force the conclusion – on the basis of current knowledge regarding shock deformation – that the Pretoria Saltpan is the product of a meteorite impact event.'

T.C. Partridge, W.U. Reimold, S.J. Kerr and I.G. Stanistreet (1991)

Figure 24: Crater Lake, in Oregon, USA, which Wagner might have had in mind in 1922 for the origin of the Tswaing feature. Crater Lake is a sunken caldera, where a lake has formed in the depression created by the collapse of a plug of strata into an underlying magma chamber during the evolution of a volcano. In the foreground is another, younger volcanic cone emerging above the waters of the lake.

The Pretoria Saltpan (now known as the Tswaing Crater, see Plate 11) is a near-circular feature – a crater with a raised rim reaching an elevation of 119 metres above the crater floor, a rim-to-rim diameter of 1.13 kilometres, and a shallow central lake highly charged with soluble salts. It lies some 40 kilometres north-northwest of Pretoria (Tshwane) and was 'the most important salt-lick in the central Transvaal and, as such, the resort of great herds of game' long before it became a source of salt for farm and domestic consumption (Levin, 1991).

The crater lies entirely within the Red Bushveld Granite that overlies the layered rocks of the Bushveld Complex, and the encircling rim of low hills is built of a fragmental rock (a breccia) of granite debris, dipping away from the centre and spreading over the surrounding terrane. Exploratory drilling has shown that approximately 60 metres of sands, gravels and granitic breccias lie between the underlying foundation of fractured but coherent granite, and the overlying sequence of carbonate-rich and saline muds.

It was G.A.F. Molengraaff's belief (1904: 57) that the Pretoria Saltpan was a manifestation of violent volcanic activity during the closing stages of emplacement of the Bushveld Complex. Wagner's (1922) conclusion that it was a volcanic crater with the structure of a sunken caldera resulting from the collapse of a central plug or cylinder into an underlying magma chamber was in part based on his comparison of the Pretoria Saltpan with a very similar feature in Germany – long before the claim that the latter was itself an impact crater, originating in much the same way as the famous Meteor Crater of Arizona. (Meteor Crater is some 200 m in depth and 1.2 km in diameter, showing a raised rim and an encircling blanket of debris ejected from the depression – therein displaying a striking resemblance to the Saltpan.)

Interpretation of the evidence at the Pretoria site was in earlier days clouded by the occurrence of volcanic rocks that suggested a volcanic origin for the feature. Against this, however, a later study by Partridge and colleagues (1991) revealed features generally associated with the massive shock that would result from an impacting meteorite: the deformation or destruction of the characteristic, ordered crystal structure of minerals such as quartz, micas and feldspars; the presence of glass spherules and droplets; metal fragments; and coatings of glass on breccia fragments. On the evidence of carbon-14 dating, these investigators estimated that infilling of the crater by sands and muds had proceeded at the rate of about 1 metre per 2 000 years, giving the time of the impact as around 200 000 years ago. This is some four orders of magnitude (10 000 times) younger than the age implied by Molengraaff's reconstruction 90 years earlier.

A considerable literature on the subject developed with further research in the years following 1991, led by Professor Uwe Reimold at the University of the Witwatersrand, and summarised in the handbook, *The Tswaing Meteor Crater*, as the site is now known (Reimold et al., 1999). The estimate of age based on carbon-14 dating was confirmed by a method known as fission-track dating, which returned an age of 220 000 years for the formation of the glass particles. The volcanic rocks that had misled earlier investigators were dated to a much earlier time – 1 300 million years – and their occurrence here may now be regarded as no more than a red herring, a mere coincidence. Analysis of small particles of debris from the site has revealed strong contamination by certain elements – nickel, chromium and platinum-group metals – that would be expected from impact with a meteorite. It has been calculated that the missile would have been some 30 to 50 metres in diameter and had impacted at a velocity of 20 to 30 kilometres per second, releasing the same amount of energy as 100 Hiroshima-type bombs.

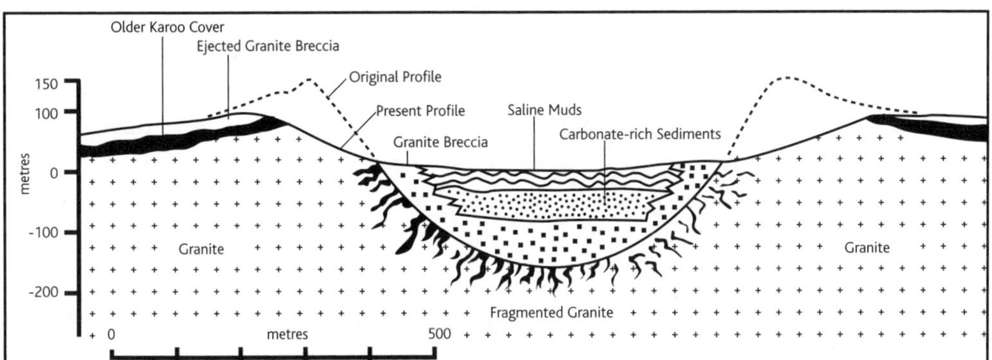

Figure 25: The structure of the Tswaing Meteor Crater as interpreted by Reimold et al. (1999).

16
Schoenmakerskop quartz porphyry

' "Courage!" he said and pointed toward the land,
This mounting wave will roll us shorewards soon.'

Lord Alfred Tennyson (1809–92), in 'The Lotos Eaters' (1832)

To the west of Port Elizabeth, between Schoenmakerskop and Sardinia Bay, is a gully whose shoreline is littered with rounded, water-worn pebbles of a greenish quartz porphyry (an intrusive igneous rock resembling fine-grained granite). For many years this has been an enigma, as the occurrence could be likened to an extra, unrelated piece found in a jigsaw puzzle. The pebbles lie scattered around outcrops of the Table Mountain Sandstone, and the occurrence of intrusions of quartz porphyry into this formation would be quite anomalous in terms of what is known about the history of this group of rocks elsewhere.

The riddle would appear to have been solved by the discovery of the wreck of a Portuguese ship, the *Sacramento*, which went aground 14.5 kilometres east of Cape Receife (near Port Elizabeth) and broke up in 1647 while on its maiden voyage from Goa to Portugal. The survivors of this wreck, together with those from the *Atalaya* (another vessel that had foundered off the Fish River mouth), eventually reached Inhaca Island – after being rescued by a Portuguese galley while the party was attempting the hazardous trek on foot towards Lourenco Marques. Only later in 1648 did the survivors finally reach the safety of Goa once more.

A detailed account of their salvage operations, in shallow waters off Cape Receife, of 40 cannons forged long ago in the then-famous Baccaro foundry in Macao, is given by Allen and Allen (1978), the discoverers. The seemingly obvious conclusion to draw from this would be that the quartz-porphyry pebbles had been loaded as ballast (to help ensure stability of the ship) on board the *Sacramento*. Subsequent reports in the press then claimed that a rock identical to that of which the pebbles are formed had been traced back to the original quarry from which they came, at a locality in Portugal.

All of this would seem to answer the geological riddle over the provenance of the pebbles, but there remain some curious anomalies in the reconstruction of the story. The *Sacramento* had been on its maiden voyage, which begs the question how ballast from Portugal came to be taken on board in distant Goa. Furthermore, the vessel had been very heavily laden with iron and brass cannons in the holds, so it could be asked whether ballast, so much lighter and bulkier than metal, had even been necessary? Or, had the rock merely been used to adjust the trim of the vessel? The answers to these questions could well be straightforward, but they do point to the need for further historical research.

PART THREE

Fossils and the Beginnings of Life

17
Introduction: Fossils and ancient life

'It cannot be doubted that the geological record, viewed as a whole, is extremely imperfect . . .

. . . we have no right to expect to find in our geological formations, an infinite number of . . . transitional forms, which . . . have connected all the past and present species of the same group into one long and branching chain of life.'

Charles Darwin (1859: 229, 235)

Fanciful ideas about the origin of fossils were, in earlier centuries, about as bizarre as were some of the creatures and plants themselves found entombed in rock. There were endless arguments about what these mysterious objects were. Spawn or germs that grew there? Failed experiments discarded by the Creator, or merely stones of a particular shape? Objects 'placed there by the devil, in order to deceive . . . mankind' (Geikie, 1905: 46)? These were some of the interpretations that surfaced.

For sheer imagination, it would be hard to outdo the Victorian, John Ruskin, for the fantasy in his *Seven Lamps of Architecture* (1880). Ruskin was a wealthy and widely travelled critic of art and architecture, with a passion for stone and a reputation as a profound thinker, but some of his passages were, to say the least, distinctly eccentric. He was certainly out of his depth in palaeontology. He wrote:

> No reason can be given, I believe, why the system of the higher animals should not have been capable . . . of secreting flint, instead of phosphate of lime, or, more naturally still, carbon, so framing the bones of adamant [diamond] at once. The elephant or rhinoceros, had the earthy part of their bones been made of diamond, might have been as agile and light as grasshoppers . . . (1880: 45)

Old bones changed to stony material are common enough, but the mind boggles at his image of skeletons constructed of crystalline diamond, and of woolly mammoths and rhinos performing pirouettes in the air. Ruskin had a fertile imagination, but he certainly had no concept of the massive pressures and high temperatures needed to crystallise carbon as diamond. Had the carbon that he wished to see forming the 'bones' of elephants crystallised in its alternative natural form as the mineral graphite, his skeletons would have possessed strength little better than that of soap, and his grasshopper-like rhinoceros would have had all the agility of a giant amoeba.

So, what exactly is a fossil? A simplified definition would be: a manifestation or part of a once-living creature or plant, or something formed by the activities of such an organism, with the implication that it is preserved in sedimentary rock, amber or hardened tree resin, layer of coal, or ancient natural tar pit, etc. A distinction is drawn between true fossils, representing the remains or impression of the organism itself (be it shell, bone, woody tissue or a film of carbon, or a chemical replacement of the hard parts by other minerals), and trace fossils or *ichnofossils*, which are structures created by the organism during its lifetime (such as burrows, footprints, etc.).

Fossils range in size from the remains of gigantic 30-metre dinosaurs found in the Gobi Desert of Mongolia, to marine trilobites a few millimetres in length, and from there through foraminifera and radiolaria of pinhead size, down to fossil bacteria seen only with the aid of optical or electron microscopes. That portion of geological time during which fossilised material was preserved has traditionally been termed the Phanerozoic – the time elapsed since the start of the Cambrian Period (~540 million years ago) – but the term has lost precision since more and more imprints of soft-bodied, primitive animals began to turn up in still older Precambrian rocks.

FOSSILS AS SNAPSHOTS OF THE PAST

Fossils require rather special conditions for their preservation, and only a minute fraction of the animals and plants of the past are preserved today. Rapid burial after death is essential, before decay and disintegration set in to destroy or scatter the remains, and it is usually sediment settling or precipitated from water, or fine volcanic ash, that provide the best media for preservation. Lakes, ponds, estuaries, the oceans or ancient river courses are the most productive environments.

It would seem a long shot to link the accumulation of fossil remains also with the violence of volcanic activity, but craters blown out long ago by the explosive emplacement of kimberlite pipes provided a niche for thriving plant and animal life in the Kenhardt region of the Northern Cape. Roger Smith (1995) of the South African Museum has reconstructed a picture of craters that filled with rainwater into which volcanic ashes slumped at Stompoor, some 70 million years ago, close to the end of the Cretaceous Period. Plants that flourished there became a haven for fish, frogs, snails, mussels and tiny crustaceans known as ostracods, as well as small dinosaurs the size of sparrows – now fossilised in the muds that collected there. Likewise, the moulds that hardened around the bodies of the hapless citizens of Pompeii, who died in agony in the hot volcanic ashes that spewed from Mt. Vesuvius in 79 AD, are fossils in no less a sense than are the remains of dinosaurs.

Under favourable conditions, the abundance of fossilised remains is astonishing. The Burgess Shale band of marine origin in the Canadian Rockies yielded 65 000 specimens of trilobites, sea urchins, molluscs

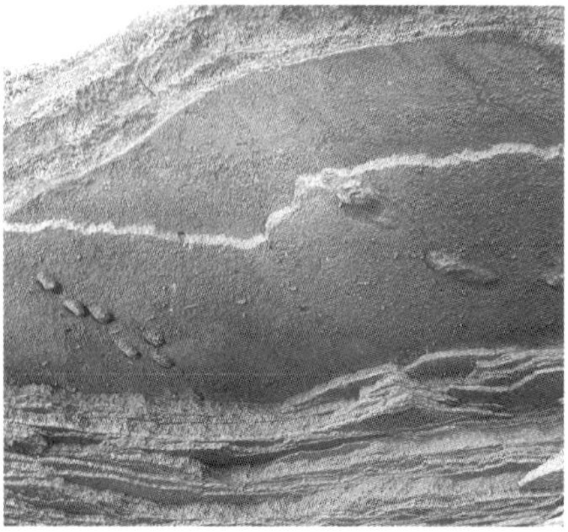

Figure 26: Human footprints (right) and those of a dog or hyena (left) preserved in calcareous sandstone at Nahoon Point, East London, dated to 29 000 years. Photograph, looking up at the under-surface of an overhang, taken by Professor E.D. Mountain in 1972, shortly before the outcrop collapsed or was vandalised.

and worms from a small outcrop. Lyell (1865: 25) recorded a study by a Professor Ehrenberg of Berlin, who claimed that a siliceous rock, tripoli, of Tertiary age in Bohemia, contained the remains of up to 40 000 million minute plants (diatoms) per cubic inch.

In general terms, however, the fossil record is woefully incomplete. Great gaps in it were the cause of bitter fights between those who believed in continuous evolution of species (as did Jean-Baptiste Lamarck, 1711–1829, long before Charles Darwin, 1809–82) and those who invoked periodic extinction by catastrophic events and spontaneous appearance of new species (as did Georges Cuvier, 1769–1832, the revered French anatomist).

Trace fossils come in a variety of forms. Human footprints have been found in hardened wind-blown sands at Saldanha Bay, and at Nahoon on the coast near East London (Roberts, 1998) where carbon-14 dating gave an age of approximately 29 000 years. Together with the footprints at Nahoon are prints of bird claws and the paws of either a hyena or a domesticated dog

Figure 27: ***Octochara crassa***, one of the world's oldest reconstructible charophyte algae, found by Robert Gess in the Grahamstown fossil site. The eight-rayed whorls (2 cm in diameter) of 'leaflets' were stacked at intervals up fine stems supported by the water in which they grew. Charophytes are believed by many to represent the link between simple algae and more complex higher plants.

Figure 28: One of the eight species of Upper Devonian fish new to science recovered from the Grahamstown site by Robert Gess in 1985. Named ***Diplacanthus acus***, this is the first example of this genus of ***Acanthodian*** fish found in the southern hemisphere.

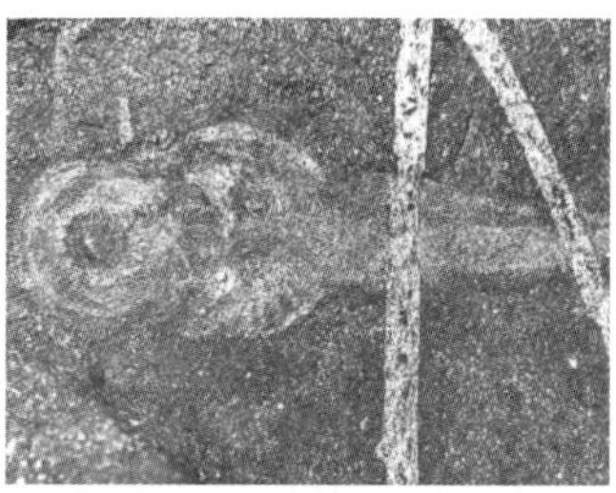

Figure 29: The most ancient known fossil lamprey – a primitive, parasitic fish without jaws, but possessing a sucker disc (left end of photograph) and teeth – 35 million years older than any other known specimens. Full length of this fragile specimen, recovered from the 360-million-year-old Grahamstown site, is 4 cm. (Photographs courtesy of Robert Gess.)

(see Figure 26). The ages of these indications of the presence of ancient humans pale into insignificance when compared with *Australopithecine* footprints in Tanzania dating back 3.6 million years, and others in Kenya dated at ~1.5 million years.

Dinosaur footprints have been discovered in many parts of the world, and there has recently been an uncovering of tracks left by a herd of the mammal-like reptiles, Dicynodonts, on the farm 'Sante Sana', 45 kilometres east of Graaff-Reinet in the Eastern Cape (see Plate 12). Here Dr W.J. de Klerk (2002) of the Albany Museum has described 158 prints left by a species about the size of a very large pig, within an exposure covering some 40 square metres. The length of stride, the depth of the impressions, and the shapes of the footprints permit the identification of the animal (*Aulacephalodon*) that left them in ancient muds some 253 million years ago.

FOSSIL FINDS AS GUIDES TO THE ECOLOGY OF THE PAST

'Gold is where you find it' was the motto of the early prospectors. This could equally well be said of fossils, which have a habit of turning up in the most unpromising terranes. The hard, bluish quartzites of the Witteberg Group that build the hills in the Albany District had been written off many years ago as of scant interest, apart from plant debris in some thin shale bands, and rough impressions of plant stems in the coarse-textured sandstones. Hatch and Corstorphine (1905: 88) had dismissed the strata as 'almost as barren as the Table Mountain Series', while Du Toit (1954: 261) concluded that 'little weight can therefore be placed on the older records' of what had been found.

This jaundiced view changed quite dramatically in 1985 when a new road was cut through a high ridge south of the city of Grahamstown. As the road-cut deepened, there emerged a band, up to 6 metres thick, of black, carbonaceous shale bearing an extraordinary wealth of fossil material, the presence of which had not been suspected despite the energy of numbers of fossilists since the times of W.G. Atherstone in the 1850s. Postgraduate student, Robert Gess, and his supervisor, Dr Norton Hiller, subsequently collected a wealth of material. With hammer and chisel, and a penknife for gently prizing apart the laminae of the fragile shale, Gess extracted close on 1 000 specimens belonging to no less than 50 different algal, plant, arthropod and fish taxa (taxonomic groups), and the catalogue that recorded their findings (Gess & Hiller, 1995) ran to 80 pages of published text.

After several years of work in the exposure, which left him looking like a coal miner at the end of each day, Gess summarised his findings in the words: 'Due to the unique nature of the site, the majority of specimens collected belong to formerly unrecorded species' (2002: 58). Of outstanding scientific importance were two new genera of algae that yielded 'four of the world's oldest reconstructible charophyte algal species'. A Progymnosperm of tree-like form, *Archaeopteris*, was the

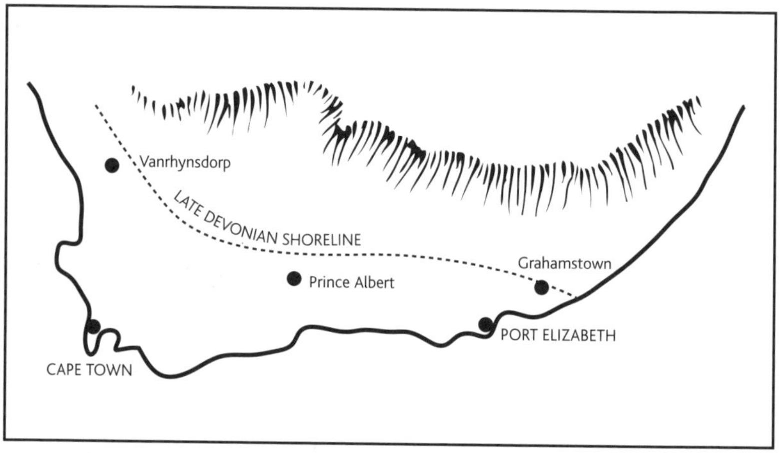

Figure 30: The ancient shoreline along which the sea lapped our southern coast during Late Devonian times, when the Witteberg Group sediments were laid down. Map simplified after that presented by Gess (1995).

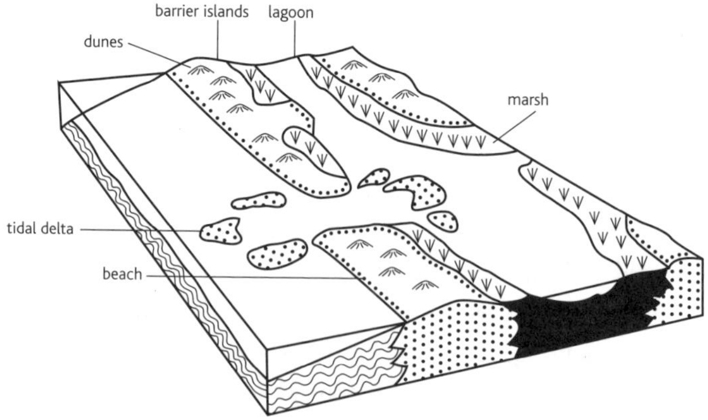

Figure 31: The manner in which the fossil-bearing shales (shown in black) were deposited in a sheltered coastal-lagoon environment in Upper Devonian times, and protected on the seaward side by barrier islands, as reconstructed by Gess and Hiller (1995) and Gess (2002). The exposure now lies some 50 kilometres from the present coast.

Figure 32: The sands deposited on the sea floor during the Late Devonian are today represented by Witteberg Group quartzites. Intense folding of these into vertical attitudes within the Cape Fold Belt is here displayed at Lake Mentz in the Eastern Cape. (Author's photograph reproduced here with permission of S.A. Council for Geoscience.)

first recorded in Africa. No less than eight species of fish were recovered, and 'probably all belong to hitherto unknown species'. All of this was unsuspected and hidden until the bulldozers moved in, and was unearthed within sight of a hundred-year-old university, and a famous and even older Albany Museum.

The drive through the mountainous terrane of the southern Cape province carries us through deep gorges cut into resistant ridges forming the high ground. Even the most imaginative would have difficulty in visualising the nature of this terrane in Late Devonian times, when the sea lapped South Africa along its southern coast (see Figure 30). The work of Hiller and Gess has now painted a picture of this coastline (see Figure 31) at that ancient time. The black, fossil-bearing shales are identifiable as sediments laid down in a lagoon protected from the open sea by sandbars and islands, between which the tides raced at both ebb and full flood. The fossiliferous black shales represent the stagnant, brackish part of the ancient lagoon, and the setting would not have been very different to that of the numerous lagoons now dotted along our coast. With a subsequent change in sea level, there would have been a return to dominantly deltaic conditions that spread cleaner sands over the top of the fossil-bearing strata, and hid them from view for the next 360 million years.

Further examples revealing the ecology of our land in the past are provided by our Karoo Sequence. The nature of the vegetation during the laying down of the Ecca Group was revealed by a fossil forest that emerged from the dried-up bed of the Vaal River near Vereeniging during a time of drought in 1906 (Mellor & Leslie, 1906). Tree stems identified as *Noeggerathiopsis*, estimated to have reached heights of 30 metres, with their roots spreading out over areas 13 metres across in the underlying coal seams, gave a good indication of how the coal seam itself had been formed by the accumulation of vegetable matter.

By contrast, the change to an arid climate at the end of Karoo times is confirmed by fossil termite nests preserved in desert sands of the Early Jurassic Period (Clarens Formation) close to where the international borders between South Africa, Zimbabwe and Botswana meet (Bordy et al., 2004). These nests form clusters of sandstone pillars, elliptical in cross-section, with preserved heights of up to 3 metres. Internally they display a network of anastomosing (intercommunicating) tubes, passages, vertical shafts and open spaces between the interiors and exteriors of the pillars. Mapping has established a clearly defined preferred orientation of the ellipses, corresponding to 10° to 20° East during the Early Jurassic. This suggests that the mound architecture was controlled by the angle of the sun, and designed to create optimal climatic control within the nest. The conclusion here is that the nest structures are consistent with the semi-arid environment indicated by the nature of the sediments themselves, and point to extreme diurnal variations of temperature, with the nests having been constructed within seasonally flooded drainage hollows.

FOSSILS OF DUBIOUS ORIGIN: FARCES AND HOAXES

The origin of trace fossils is not always obvious. *Spirophyton (Zoophycos)*, found in the quartzites of the Witteberg Group, is a curious structure showing curved striations radiating in a swirling pattern from a central pit that may be several centimetres deep. In older literature this was attributed to a purely inorganic origin, while Schwarz (1912: 146) termed it a seaweed of screw-like form. Du Toit (1954: 261) interpreted it as the burrow of a worm, *Taonurus*. The swirl of ribs spirals sometimes in an anticlockwise direction, and sometimes clockwise – an observation that prompted one collector to quip that some worms must have been right-handed while others were left-handed, or ambidextrous.

Another category of structures (pseudo-fossils) is often mistaken for fossils. *Dendrites* are dark films of manganese or iron oxides on rock surfaces into which groundwater has seeped along cracks, and deposited stains bearing a striking resemblance to the fronds of plants, particularly ferns (see Plate 13). *Nodules* and *concretions* form by crystallisation of mineral matter (often calcium carbonate) in ball-like or ovoid forms within other host rocks (commonly porous sandstones or mudstones). *Septarian nodules* are those traversed by cracks infilled by further deposition of mineral matter, usually of a different colour, to produce a structure that may superficially resemble the fossilised carapace of a tortoise.

In nature there are also a number of tube-like structures that might or might not be fossils. In very young limestones associated with coastal dunes hardened into rock along our coasts, there are near-vertical tubes which are the cavities left behind by the roots of plants that died and rotted there. Quite different to this are *fulgurites*, which are tubes of sand fused to glass, produced by lightning strikes onto desert sands.

'I've dug up fragments of an ancient spacecraft in my lands – perhaps from another planet!' was once phoned in by an excited farmer near Fort Beaufort. This proved to be a sandstone traversed by two sets of natural fractures angled at ~30 degrees to each other, with the cracks infilled by red iron oxides. Where the sandstone matrix had crumbled away, a framework was left behind that certainly did bear a striking resemblance to a boxwork of rusted steel plates, but which was in reality a prosaic example of *snuffbox weathering*.

Fossil hunting is not just about dry bones and dust. Like any other science, from time to time it offers up elements of pure farce. A bizarre case involved the eccentric Oxford academic, Dr Robert Plot, keeper of the Ashmolean Museum. He was a firm believer in the former existence of human giants reaching 3 metres in height, and one bony fragment that he discovered in 1676 belonged, he declared, to such an individual. This was subsequently given the wildly implausible name of *Scrotum humanum*, but it was actually the femur (thigh bone) of a dinosaur (McGowan, 2003: 1).

An echo of this surfaced in a South African first-year university examination paper asserting that 'crinoids are marine animals that lie on their backs on the sea floor, waving their testicles in the ocean currents to gather food'. This evoked from the bemused examiner the comment that if the candidate had been clearer on the spelling of 'tentacle', this would not have been a bad effort. (Crinoids are marine animals with a fossil record dating back to the Early Palaeozoic. They belong within the Phylum Echinodermata, which also includes the sea urchins and starfish, and are characterised by a cup-shaped body, five flexible arms and an anchoring stem.)

Perhaps the most famous scientific hoax of all time – the Piltdown skull – deserves brief mention here, for it fooled the leading anthropologists of the day for more than 40 years. During 1908–12 an English lawyer and amateur naturalist, Charles Dawson, claimed to have discovered skull fragments in a gravel pit at Piltdown, Sussex. From these, Sir Arthur Smith-Woodward of the British Museum reconstructed a skull that threw the anthropological community into turmoil, for the brain-case appeared to be indisputably human, while the jaw was indistinguishable from that of a young chimpanzee. Was this the long-sought 'missing link' between man and ape? Then, from a second skull reputedly unearthed a few kilometres away, Sir Arthur Keith, the anthropologist, carried out a reconstruction that yielded a human-like face.

The realisation that this was an elaborate hoax came slowly, following on the work of Dr Kenneth Oakley, at the Natural History Museum, on the absorption of fluorine by old bones buried in the earth. The amount that is absorbed is governed by the length of time that bones have been buried – any differences shown by analysis of separate fragments from the same deposit should therefore indicate whether a match, in terms of age, is valid or not. Oakley's work suggested that the bones were in the region of 50 000 years old, as opposed to previous estimates of up to 1 million years.

Just at the time when doubt was turning to certainty that this was one of the most successful scientific hoaxes ever perpetrated, Ruth Moore wrote her book, *Man, Time and Fossils* (1954). In her main text she noted that tests up to that date had suggested that the 'jaws and cranium could have belonged to the same being', but in a footnote added 'Oakley emphasizes that it is still possible that the remains represent two creatures, though this does not now seem likely' (1954: 316). A note apparently added to Moore's manuscript in December 1953, then included the results of a re-investigation, revealing that the jaw and teeth undoubtedly belonged to a ten-year-old chimpanzee, filed down mechanically to present seemingly human traits, and then stained with chemicals to give the appearance of age (1954: 321–23). What has never been resolved is the identity of the hoaxer. At various times there has been intriguing discussion on who that might have been, with fingers pointed towards some surprising names – Dawson, Smith-Woodward, Oakley and Arthur Conan Doyle. Or was it someone else?

The intense rivalry between Edward Cope (Professor at the University of Pennsylvania) and O.C. Marsh of Yale University in the late nineteenth century drove them to a competitive frenzy of collecting that yielded thousands of fossils. Given that they would have been unable to contend with the demands of cataloguing and reconstructing the tons of bones gathered, mistakes were inevitable. Cope never lived down the gaffe of placing the head of *Elasmosaurus* on its tail – an error that Marsh exploited to the full in later acrimonious confrontations.

There also was the farcical case of a magnificent, near-complete reptilian fossil skeleton recovered from Karoo strata, that was to be placed on display in the main concourse of a South African university building. When the glass case built by a cabinetmaker was delivered, it was found to be slightly too short to accommodate the fossilised remains at full stretch. The decision was made to 'squeeze it up a bit' inside the cabinet, rather than to embark on an expensive modification. During a subsequent visit by an eminent vertebrate palaeontologist, he was heard, after long scrutiny of the humped back, to comment drily that this creature had seemingly expired from either chronic constipation, or a strangulated hernia of some sort.

Then there was a sea urchin that might have found a place in literature. The scene in the Court of Justice of Venice is, perhaps, one of the best known in all the works of Shakespeare. In the *Merchant of Venice* ACT IV, Scene I, we read:

> Portia: Therefore prepare thee to cut off the flesh.
> Shed thou no blood; nor cut thou less, nor more,
> But just a pound of flesh: if thou tak'st more,
> Or less, than just a pound, be it so much
> As makes it light or heavy in the substance,
> Or the division of the twentieth part
> Of one poor scruple, nay, if the scale do turn
> But in the estimation of a hair . . .

Here was the moneylender, Shylock, demanding that a pound be cut from the living flesh of Antonio. He is thwarted by Portia demanding that this horrific penalty be imposed on him (Shylock) in turn if there is the slightest departure from the precise weight of one pound.

At first glance, there would seem to be no possible connection with the stony world of geology, but there is a tenuous link. What is a pound? More particularly, how was it defined in the sixteenth century?

In Simon Winchester's *The Map that Changed the World* (2001: 33–36) is a passage from which any astute lawyer would realise that Shylock had missed a trick. In Oxfordshire the trade in dairy products was in earlier years based on the 'long pound', measured by the weight of the fossil echinoid, *Clypeus ploti*, otherwise known as the Chedworth Bun, which occurs in abundance in the Jurassic sediments exposed in local quarries. The weight of these fossils is, according to Winchester, consistently about 22 ounces (624 g), rather than the normal measure of 16 ounces (454 g) to the pound. The variability in size to be expected between one fossil specimen and another, and the range between the weight of the conventional pound and the long pound, could surely have given Shylock more leeway than he imagined?

18
Early ideas about the origin of fossils

'Egyptian priests were aware, not only that the soils beneath the plains of the Nile, but also that the hills bounding the great valley, contained marine shells; and Herodotus [*c*.484–425 BC] inferred from these facts, that all lower Egypt, and even the high lands above Memphis had been covered by the sea.'

Sir Charles Lyell (1850: 7)

'They tell us that these shells were formed in the hills by the influence of the stars; but I ask where in the hills are the stars now forming shells of distinct ages and species?'

Leonardo da Vinci (1452–1519) in Lyell (1850: 22)

'. . . a certain "materia pinguis" or "fatty matter", set into fermentation by heat, gave birth to fossil organic shapes . . .'

Agricola (1494–1555), mining engineer and metallurgist in Lyell (1850: 23)

'. . . they are mere stones, which had assumed their peculiar configuration from the influence of the heavenly bodies . . .'

Mercati (1574) in Lyell (1850: 24)

'. . . shells and submarine matter of the Veronese, and other districts, might have been cast up upon the land by volcanic explosions . . .'

Majoli (1597) in Lyell (1850: 24)

'"... there is no such matter as Petrifying of Shells in the business." The "Cockle-like stones" were stones and had always been stones.'
Cutler (2003: 134) on the written works of Lister (1638–1712)

'... partly owing to fish-spawn received into the chinks of the earth in the water of the Deluge.'
Edward Lhuyd (1660–1709), Keeper of Geology, Ashmolean Museum, Oxford, as quoted by Sir Archibald Geikie (1905: 78)

'Lang (1670–1741) assigned their origin to dust-like germs in the earth.'
Orlov (1962: 10)

'I have been told that even within the last century a learned divine of the University of Oxford ... [maintained] that the fossils in the rocks had been purposely placed there by the devil, in order to deceive, mislead and perplex mankind.'
Sir Archibald Geikie (1905: 46)

Amongst the many strange explanations for the origin of fossils – unspecified 'plastic forces', influence of the stars, fermentation, germs or spawn in the Earth, evaporation, jokes played by nature or the devil – there were some that came very close to the truth. But Church dogma had held science on a tight rein, as is shown by the great number of sixteenth– to eighteenth–century works that refer the origin of fossils back to the Biblical Flood.

The inimitable Leonardo da Vinci (1452–1519), genius and supreme master in many fields – painting, sculpture, engineering, and anatomy– had also found time to draw some incisive conclusions about marine fossils that had been found at sites 600 metres above sea level in Italy. His essay, 'Seashells in the Mountains' (1480) reprinted by Bolles (2000: 105–111) shows how Da Vinci began with the conventional premise that these were merely detritus left by the Biblical Flood, but over a period of some 25 years came to assemble arguments in favour of the fossil bivalves, gastropods, corals and fish being the petrified remains of marine animals that had died and accumulated in the layers where they are now found.

Da Vinci's argument was that a rising flood, as massive as that indicated in Biblical works, would have scattered such remains indiscriminately at all levels over the mountain slopes, rather than gather them in distinct rock layers at specific levels. From there it was a relatively small step, without actually challenging the Flood legend, for him to question whether it could have been as high and universal as claimed, and to ask to where a mass of water as vast as this could have disappeared. This question was still being asked several centuries later, when geologists found fossil shells at an elevation exceeding 5 500 metres in the Himalayas. The most remarkable feature of Da Vinci's essay is that it was written at a time when it was dangerous even to commit such thoughts to paper.

The work of the Dane, Nicolas Steno (1638–86) was another milestone that led to the conclusion, against prevailing dogma, that all fossils are the remains of living organisms, but he still remained tied to the concept that they were an aftermath of the Flood. His dissection of the head of a massive Great White shark (Cutler, 2003: 53) that had been dragged ashore by fishermen revealed the true origin of the fossilised teeth (*glossopetrae* or 'tongue stones') that had been objects of wonder at that time, and were regarded as having medicinal powers.

The grip of the Church on all ideas that might impinge on Creationist doctrine is further illustrated by the demand that the Comte de Buffon recant in 1769 even innocent remarks such as 'the waters of the sea have produced the mountains and valleys of the land'. He capitulated abjectly with: 'I declare that I had no intention to contradict the text of Scripture . . . and I abandon everything in my book respecting the formation of the earth, and, generally, all which may be contrary to the narration of Moses' (see Lyell, 1850: 41–42).

FOSSILS AS PAGES IN THE GEOLOGICAL CALENDAR

Only in the early nineteenth century did palaeontology move, even if tentatively so, onto firm ground, following the publication of major works by Cuvier (1812) and Lamarck (1815) in France, and William Smith (1815) in England. Cuvier (1769–1832) had used fossils to identify individual beds within the Paris Basin, but remained rigidly opposed to the concept of evolution, believing that each species was periodically extinguished by catastrophe, and then replaced by others.

The story of William Smith (b.1769) as told by Simon Winchester (2001) reveals a remarkable man who made the giant step in Britain towards recognising the value of fossils as a scientific tool. Smith was born the son of a blacksmith in Oxfordshire and became a self-taught surveyor in coal mines, and an engineer in canal-building. Here his interest in the infant science of geology was kindled by what the canal-digging revealed, and he began to lay the foundations of stratigraphy – establishing the sequence, distribution and relative ages of rock strata. As early as 1795 came the insight that individual beds could be recognised and correlated over long distances by the marine fossils entombed within them. Beds of different ages would carry different fossil assemblages.

Out of this discovery evolved Smith's (the first) geological map of England, Wales and parts of Scotland, published in 1815 – an immense task accomplished without help, and with quite trivial financial reward. Tragically, it did not bring him immediate acclaim, for in the succeeding years he had had to endure blatant plagiarism, if not downright theft, of his great work, denigration by the scientific establishment of the day, and even incarceration in a debtors' prison before formal recognition of his work came. This was in the form of the first award of the Wollaston Medal in 1831 by the Geological Society of London – an honour that came sixteen years late.

We might ponder how thoroughly the rigid distinctions of 'class' permeated the scientific community in the nineteenth century, to separate a wealthy and well-educated scientist (Lyell) from someone who had risen from a level close to that of labourer. The name 'William Smith' does not even appear in the detailed index to the 794 pages of Lyell's *Elements of Geology* published in 1865. Others were more generous – the Oxford Professor, W.J. Sollas (1905), grants Smith 30 pages of laudatory text in *The Age of the Earth*, and Smith's place in history is now secure.

The truth that William Smith had revealed was not universally accepted at first. The formidable Thomas Huxley, FRS (1825–95), had gone so far as to argue that the occurrence of the same fossils in strata that lay geographically remote from each other was in itself proof that they could not be of the same age. His argument was that every species of plant or animal had evolved at some specific place and time on the Earth's surface, and had slowly spread outwards from there. Therefore, as generation followed generation, their descendants should eventually become entombed in sediments that lay progressively further away from their place of origination. This was subtle reasoning indeed, and it persuaded many that the word 'contemporaneous' should be dropped in favour of the term 'homotaxial' for strata in which the fossil assemblages matched.

Sollas (1905: 266–73) made short work of Huxley's contention. He pointed out that Huxley's logic would lead us to expect actual reversals of the normal sequence of fossil-bearing zones at some places. Sollas's rebuttal drew upon evidence provided by the fossil ammonites, animals with a shell that displays strong evolutionary features. Certain groups of these ammonites found in Jurassic rocks of India occur in such abundance as to suggest that they originated there, while other groups that are more abundant in Europe apparently originated within a different 'Atlantic' province. Each assemblage, however, exhibits some admixture with groups from the other province, showing that migration of species between the two regions had indeed occurred. If Huxley's argument that migration from one region to another required a significant passage of time were to be accepted, the occurrence of Atlantic-type ammonites in Indian strata would mean that the Indian strata are younger in age than the homotaxial European strata. This inference, however, leads straight into a paradox – the occurrence of Indian groups in European strata would make the European strata the younger. As Sollas put it, the conclusion would be that 'each must be older than the other, which is impossible' (1905: 278).

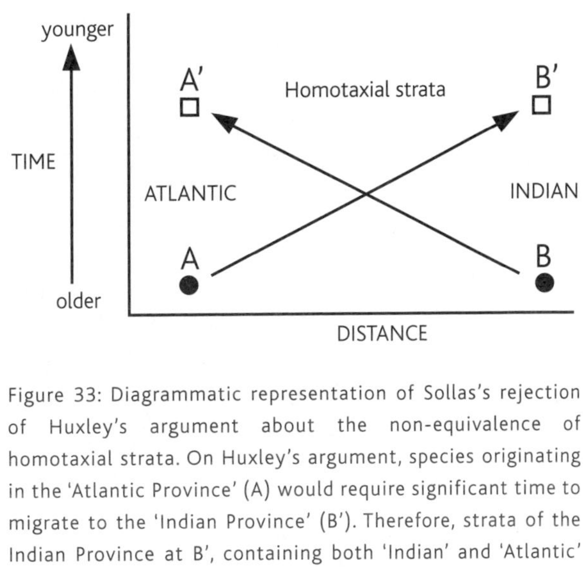

Figure 33: Diagrammatic representation of Sollas's rejection of Huxley's argument about the non-equivalence of homotaxial strata. On Huxley's argument, species originating in the 'Atlantic Province' (A) would require significant time to migrate to the 'Indian Province' (B'). Therefore, strata of the Indian Province at B', containing both 'Indian' and 'Atlantic' species, should be the younger. Paradoxically, by the same reasoning, 'Atlantic' strata at A', containing some 'Indian' species, should be younger than those at B, thereby indicating the Atlantic Province as the younger.

It had already been established that precisely the same sequence of ammonite-bearing zones of Jurassic age is to be found across Europe, India, Argentina and Chile. Likewise, strata containing the extinct fossil graptolites of Palaeozoic age (animals distantly related to the 'Portuguese man-of-war') display precisely the same zonal sequence in Scotland, Wales, France, Spain,

Germany, Scandinavia, North America and Australia. Inversions of the normal sequence to be expected from the reasoning of Huxley are simply never found. The significant point here is that both the ammonites (free swimmers) and graptolites (attached to floating seaweed, or possessing a swimming bladder) were capable of spreading rapidly through the oceans of the world. The time required for migration would have been vanishingly small in a geological context.

Only the most obtuse would deny the contribution made to geological science by the study of fossils through the nineteenth and twentieth centuries. The fauna and flora preserved in the rock record had provided a method for assigning relative ages to Phanerozoic rocks in separate continents, and of correlating them, long before dating by radiometric methods was developed. The slow subsidence of controversy over their origin then set the stage for controversy of a different sort – what did the evidence mean? Here were the roots of fundamental warfare on a different front, the issues now being the profound questions about the origin of life itself, and its subsequent evolution.

19
Evolution, Extinctions and the Origins of Life

'Let th' Earth bring forth Fowle living in her kind,
Cattel and Creeping things, and Beast of the Earth,
Each in their kinde. The Earth obey'd, and strait
Opn'ning her fertil Womb teem'd at a Birth
Innumerous living Creatures, perfet formes,
Limb'd and full grown . . .'
John Milton (1674, Book VII: lines 451–56)

'The Egyptians . . . taught . . . that after each great catastrophe the same species of animals were created over again.'
Sir Charles Lyell (1850: 15)

'If I had been present at the creation, I would have given some useful hints for the better arrangement of the Universe.'
Alfonso the Wise (1221–84), King of Castile (Cohen & Cohen, 1963: 3)

'One of the lessons we draw from the fossil record is that, for the most part, species do not last very long . . . more than 99 percent of all species that have ever lived are now extinct . . .

Periodic mass extinctions therefore characterize Earth history, as do the rapid recoveries from these events . . .

According to some estimates, as much as 50 percent of the world's species will be extinct within three decades.'
Richard Leakey (1992: 355, 354, 353), palaeoanthropologist

The instantaneous creation of all living species, each perfectly formed and mature, is what is implied in Milton's lines – a paraphrase of the biblical version of 'the Creation'. There was no flexibility in this biblical doctrine. Then there was the belief of the ancient Egyptians in a sequence of extinctions (the wiping out of species) followed by rebirth of the same things. The fossil record preserved in our rocks simply does not permit such a philosophy. Here we have the crux of the bitter controversies that split the geological fraternities of the eighteenth and nineteenth centuries – well told, amongst others, by Ruth Moore in *Man, Time and Fossils* (1954) and by Christopher McGowan in *The Dragon Seekers* (2003).

THE ORIGIN OF LIFE IN CHRISTIAN CHURCH DOCTRINE

Where, when and how life first appeared on Earth is possibly the most fundamental question we could ask ourselves. The biblical version tells us that all the lesser creatures appeared fully formed and mature on the fifth day of Creation, only after which '. . . God formed man of the dust of the ground, and breathed into his nostrils the breath of life . . .' (Genesis 2, 7). As the age-old fascination with our Earth matured into the new science of geology, two great figures emerged on the side of Creationism and orthodoxy – Georges Cuvier (1769–1832) in France, and William Buckland (1784–1856), the first Professor of geology at Oxford University. The basic premise of Cuvier was creation and then periodic total extinction followed by the creation of new species, which, in retrospect, was a philosophy not greatly distant from that attributed to the ancient Egyptians by Lyell.

Buckland was the unshakeable advocate of traditional Church views. As an inspiring teacher, compelling leader and man of great personal charm, his every observation was shaped by seeing the hand of the Creator in all natural phenomena. In Buckland's opinion the Biblical Flood, the story of Genesis, and even commonplace features such as the convenient placement of coal seams and deposits of metallic ores in the Earth were all part of a master plan. Where Cuvier and Buckland differed was with regard to the concept of extinction: 'The idea that any of God's creatures had failed to survive cast aspersions on His wisdom and was thereby untenable [in Buckland's view]' (McGowan, 2003: 3).

LIFE FROM EXTRATERRESTRIAL SOURCES?

How life first came about on Earth remained a highly emotive subject until the taboo was broken by scientists. It was a question belonging more in the domain of the biologist and organic chemist than of the geologist, but it is to the last that we owe the gathering of the evidence that is preserved in the rock record.

In the view of some, life may have been carried to Earth by meteorites from other planets (Nisbet & Sleep, 2001). Our early Earth was repeatedly bombarded by other bodies from space, one current view being that at 4.5 billion years ago there had been an impact so great that the energy emitted had melted our entire planet, and ejected enough molten material to form our Moon. This catastrophic event is considered, however, to have created an environment so hostile as to rule it out as a plausible time for infection of Earth by extraterrestrial micro-organisms. This would have come only later.

Meteorites impacting from other inner planets are deemed by many to be a probable mechanism for bringing primitive life forms to Earth, with Mars regarded as the most likely candidate. Venus had probably lost its water to space early in its history due to a runaway greenhouse effect (Nisbet & Sleep, 2001), whereas Mars was once probably more like Earth with free water on its surface – as indicated by the data from space probes with analytical instruments on board around the turn of the twenty-first century.

LIFE ORIGINATING BY GEOLOGICAL PROCESSES ON OUR EARTH?

The notion that life began on Earth itself is an alternative view. A famous experiment was conducted by Stanley L. Miller at the University of Chicago in 1953, from which he showed that if an energy source (electrical spark or ultraviolet radiation) acts upon mixtures of gases such as methane (marsh gas), ammonia and water vapour, a range of organic compounds including amino acids can be produced. The amino acids are the 'basic building blocks' of living matter. The possibility this raised is that the molecules essential for the evolution of life may have been developed by lightning strikes in a primitive atmosphere.

A beginning in a hydrothermal setting (where heated waters are associated with active volcanism) is favoured by many, specifically in association with 'black smokers' (see Chapter 20 in this book). Whichever scenario is closer to the truth, Nisbet and Sleep concur with many others that the best evidence points to life having been present on Earth since about 3.8 billion years ago, or even earlier.

Another approach focuses on the possible role of natural crystalline materials in organising into useful biological structures the organic molecules that are believed to have evolved in the Earth's Archaean oceans (Huntress, 2003). The theme here is that 'life appears to have arisen as a natural geochemical process of the primitive Earth'. The concept that Huntress outlines holds that certain natural crystals, such as common quartz, possessing a spiral (corkscrew-like) architecture by which their constituent atoms are linked together with left- or right-handedness, are capable of preferentially adsorbing (attaching) onto their surfaces (termed chiral surfaces) certain organic molecules which also grow in either left- or right-spiralling structural forms.*

* Such compounds display what is known as *optical activity* (meaning that they rotate the plane of vibration of plane-polarised light waves passing through them, or solutions of them). It is a matter of observation that non-biological carbon compounds are not optically active, whereas organisms build chains that are. For some reason, life employs the left-handed forms of amino acids almost exclusively (Hazen, 2005).

The spotlight is focused on a number of other minerals as well, each one capable of acting as a catalyst (activator) in promoting chemical reactions that would yield the long-chain organic molecules of living forms. A review by the President of the prestigious Mineralogical Society of America, Robert Hazen (2005), states that 'researchers on the origin of life now conclude that rocks and minerals must have played key roles in virtually every phase of life's emergence'. As he points out, the difficulty to be overcome would have been the highly diluted levels of organic molecules in the vast volume of the Earth's oceans. Some means of selection and concentration would therefore have been essential. This is where natural minerals move to centre-stage. It has been demonstrated, for example, that cleaved surfaces of some common minerals such as graphite (crystalline carbon) and molybdenite (molybdenum sulphide) strongly adsorb organic species and arrange them in precise two-dimensional patterns.**

**To help imagine this it might be useful to think of the way in which pharmacists count out pills in bulk, and traders count out pearls or marbles, by dipping into the bin a flat scoop with a pattern of shallow indentations on the upper surface. The little spheres settle in the indentations, precisely following the pattern. In an analogous manner, the atoms of the adsorbed materials may be thought of as arranging themselves on the cleaved mineral surfaces in a pattern set by the mineral itself.

Research has thrown up a wide range of further possibilities. Joseph Smith (2005) at the University of Chicago has highlighted the possible role of feldspar and zeolite minerals in adsorbing organic molecules onto their surfaces and there promoting the evolution of more complex molecules. He offers a scenario in which volcanic ash might have accumulated around ponds near an active volcano. The ashes would yield both zeolite and feldspar minerals, and within the latter group there are feldspars known as perthites in which tiny domains of potassium-rich and sodium-rich feldspar phases are intergrown on a scale so fine that an electron microscope may be needed to reveal the structure. On exposure to repeated wetting and drying along the shores of the ponds, weathering would produce a honeycomb of minute cavities (less than 1/1 000 mm in size) in the crystals which, Smith asserts, 'might provide a safe home for complex organic molecules' produced by lightning strikes in the primitive Earth atmosphere. Here a self-replicating cluster of molecules might develop and then 'float off into the pond water as the first living cell'.

Smith explores, further, the possibility that some natural geological processes may have brought about not only the development of life, but also stimulated subsequent evolutionary advance. As an example, he selects the East African Rift Valley as a region in which there was a unique combination of 'an evolving large-brain primate population and an abundant supply of biochemical nutrients from active volcanism'. The rest of the tropical world, he asserts, lacked this crucial combination where volcanic activity would have increased the supply of certain elements (for example, manganese) that accelerate the rate of natural mutation, and at the same time provide other elements (for example, phosphorus, sodium, calcium and potassium) that stimulate brain growth. In this way he links the emergence of the advanced primate brain with specific nutrients introduced into the soils by volcanic eruptions.

'All life requires a source of energy', and '. . . life emerged as a direct consequence of chemical reactions derived from the interaction of . . . fluids (crustal and/or mantle-derived) with reactive and catalytic . . . metal sulfides', is the theme by which Cody (2005) at the Carnegie Institute of Washington introduces his view of the origin of life. The combining of metals and sulphur in nature produces sulphide minerals, and some of these reactions release energy and show a marked ability to stimulate organic reactions.***

*** As one example, the combination of iron monosulphide and hydrogen sulphide would drive a reaction that yields the sulphide mineral, pyrite, and organic molecules from carbon dioxide – an abundant component of our early atmosphere and oceans. Other common sulphide minerals (chalcopyrite, bornite and chalcocite) also show clear evidence of catalytic action in producing complex organic molecules.

Nearly all natural soils contain variable amounts of clay. As mineral specimens, they are hardly likely to attract much enthusiasm amongst amateur rockhounds, for they are amongst the fine-grained, sticky materials that cling to your shoes on a wet day, or are used to make bricks. In general, they represent the fine-grained breakdown products of other minerals. In a more strict sense, however, the Clay Mineral Group refers to a variety of minerals characterised on the atomic scale by a layered structure (like a pack of playing cards) and extremely small particle size. They possess the ability to expand, concertina-like, in order to accommodate water, solutions or organic compounds between their layers, and this, in the opinion of James Ferris (2005) of the Rensselaer Polytechnic Institute, New York, 'contributes to their ability to catalyze a variety of organic reactions that are critical to scenarios of life's origins'. One particular species of clay mineral – montmorillonite – that forms by alteration of volcanic ashes, shows this property in marked degree, and thereby joins a host of other minerals that show promise of having provided the pathway from prebiotic molecule to living organism.

It is a sobering thought that our most distant progenitor might thus have been a fleck of slime evolved within the dominant component of drilling mud (bentonite, containing montmorillonite) used in the boring of oil wells. This concept has a familiar ring to it. Does it bring us back full circle to the biblical stance in Genesis that 'God formed man of the dust of the ground'?

THE PROVOCATIVE CONCEPT OF EVOLUTION

While the foreshadowing of Charles Darwin's theory of evolution by Alfred Wallace, working in Malaya, is generally acknowledged, an even earlier exponent of the idea, Frenchman Chevalier de Lamarck (1744–1829), is all too often overlooked. Here was an unusually undemanding and modest man, whose professorship at the Paris Museum of Natural History did not in his own eyes elevate him above prefacing one of his publications with the words, 'I have no thought of producing a work of any importance . . .' Nevertheless, he presented a theory of evolution 'which could have upset the old order as thoroughly as did Darwin's half a century later' (Moore, 1954: 88). In short, Lamarck argued that species 'transmutated' (that is, evolved) into further species by inheriting traits developed by their forerunners during their lifetimes. The long neck of the giraffe came from stretching to reach the highest foliage. Moles lost their eyesight because they had no need of it. Aquatic birds developed webbed feet because they needed them.

Lamarck's ideas were provocative, but are now known to be unacceptable – acquired characteristics cannot be passed from parents to offspring any more than a woman who has lost an arm through injury is likely to produce a one-armed child. This is where Darwin's doctrine that all creatures vary in their attributes, and that nature favours the survival of only those whose attributes give them an advantage over others, acquired the edge over Lamarckian theory.

The concept of massive extinction of species following a rapid change to a hostile environment is something that can be readily accepted, even if the nature of the change itself remains obscure. But less well understood in the history of life is the sudden appearance of the diversity of marine animals (trilobites, sponges, coelenterates, etc.) with preservable hard parts, that proliferated in Lower Cambrian times in different regions of the world. There is no gradual build-up preceding this to be found in the fossil record. Fossils are known in rocks older than Cambrian, but these are soft-bodied animals that give little hint of the burst of evolution of animals with exoskeletons or shells, limbs and compound eyes that was soon to follow.

THE FIRST RECOGNISABLE (PRECAMBRIAN) FOSSILS

It was in the arid terrain of Namibia that the world's earliest discoveries of undoubted Precambrian fossils were made, during 1914–18, by German geologists who had collected indistinct impressions of soft-bodied organisms then new to science. These animal remains came subsequently to be known as the *Ediacaran* fauna, taking their name from a range of hills in South Australia where they were later found in 1946; today similar assemblages are known from Britain, Newfoundland, North Africa, Europe and Asia. The assemblage includes structures suggestive of jellyfish, and branching fronds that might represent primitive animals belonging to the coelenterates, but there are also impressions in rock produced by segmented worms and other animals of unknown type.

Narbonne and Gehling (2003) regard the *Ediacaran* fauna as marking 'a critical stage in the evolution of life on Earth . . .' The oldest found in Newfoundland are frond-like forms approaching 2 metres in length, and anchored to the sea floor, occurring in sediments that lie more than 1.5 kilometres beneath strata dated to 565 million years. The disappearance of this fauna is generally taken to be near the level of the first Cambrian-style shelly fossils (before 540 million years ago). The lack of preserved internal anatomical structures, and their occurrence as mere impressions in the layers of bedding planes of sedimentary strata, leaves their classification still somewhat unresolved.

C.K. (Bob) Brain of the Transvaal Museum has provided a valuable review of the *Ediacaran* fauna in Namibia (1995, 1997), and put forward the view that 'there is probably no other group of fossil organisms that has generated such a diversity of opinion, concerning their affinities . . .' One flat, oval impression with radiating ribs has alone given rise to various claims that it might be accommodated within five different phyla (the *phylum* being the highest category within which organisms having a common structural plan may be fitted) of the animal kingdom, or be a lichen within the plant kingdom, or perhaps belong within a kingdom of its own.

As stressed by Brain, the discovery of these fossil remains still leaves unanswered the question about when they originated, or what

Figure 34: Stromatolite showing nested, concentric, dome-shaped layers produced by the growth of colonies of slimy blue-green algae (cyanobacteria) preserved in Precambrian limestone of the Nama Group, Namibia, north-east of Vioolsdrift on the Orange (now Gariep) River. The scale is indicated by the matchbox.

animals may have preceded them. The decline in stromatolite fossils (structures produced by algal colonies) at 0.7 billion to 0.8 billion years ago is perhaps to be connected with widespread grazing by small animals. Genetic evidence and studies of DNA point to the origination of the most ancient animal lineages somewhere in the region of 1 billion years ago. Brain's dating of the Nama host rocks puts the upper limit of the *Ediacaran* radiation very close to the Precambrian–Cambrian boundary (between 539.4 ± 1 m.y. and 543.3 ± 1 m.y.), while evidence from Canada suggests that their time range stretched over some 55 million years.

More precise indications of the time when these primitive animals and plants evolved emerge from work by Chongyu Yin et al. (2005) at the Chinese Academy of Geological Sciences in Beijing. Dating of the rocks flooring the sequence in which the *Ediacaran* fossils occur in the Yangtze Gorges region has yielded an age of 628.3 ± 5.8 million years, suggesting an overall time range of some 85 million years for this assemblage.

MULTIPLYING SPECIES – AND CATASTROPHIC EXTINCTIONS

The astonishing variety of life forms today has persuaded many that evolution since the Cambrian Period has inevitably led to increasing diversity and numbers of species with the passage of time, but, in the view of Stephen J. Gould (1989), this is a false perspective. As he puts it: 'the history of life is a story of massive removal followed by differentiation within a few surviving stocks, not the conventional tale of steadily increasing excellence, complexity and diversity' (1989: 25).

Gould's account of the investigation of the 530-million-year-old Lower Cambrian Burgess Shale in the Rockies Mountains of Canada shows, for example, that where biologists today recognise some 20 to 30 phyla into which all modern animals can be placed, this exposure has yielded a further 15 to 20 organisms 'so unlike anything now living, that each ought to rank as a separate phylum' (1989: 99). Here were found the remains of animals of fundamentally different design that do not appear in rocks of lesser age – they simply died out. Such were the weird, five-eyed marine creature *Opabinia*, with a frontal nozzle like the hose of a vacuum cleaner, and a creature so bizarre that it was named *Hallucigenia*. This recognition of new phyla was so unexpected that it led to the famous outburst by Simon Conway Morris: 'Oh f***, not another new phylum!' (Gould, 1989: 143; Fortey, 2000: 138).

It is no simple task to uncover the trail left by living organisms as they evolved through geological time. The record is incomplete because of the patchy preservation of the remains in the first place, then subsequent destruction of fossil-bearing strata due to later geological events and, finally, imperfect exposure of what remains at the present time. Despite this, M.J. Benton (1995) has concluded in the journal *Science* that 'geological factors did not fatally damage the evolutionary signal to be read in the fossil record'. Drawing on a database of 7 186 families of microbes, algae, fungi, protists, plants and animals (both marine and non-marine) with a fossil record, he estimated the number of families in the Early Cambrian as 280, dropping to 120 in the Late Cambrian. Rapid increase to 450 followed through the succeeding Ordovician Period, to reach 600 through the Palaeozoic. Massive extinction marked the end of the Palaeozoic, but by the end of the Cretaceous the tally had reached 1 260, and 2 150 in the Pleistocene and Holocene. On these findings, Benton defined the increase in the number of all families as being exponential (that is, following a pattern of accelerating growth).

This view of exponential growth has been rejected by others (Courtillot & Gaudemer, 1996; Miller, 1998) who prefer to see successive cycles with rapid diversification at first, settling down to some steady state with the passage of time (logistic growth). The Ordovician Period, in particular, has been singled out by Webby et al. (2004) as one such a cycle – 'one of the greatest-ever diversifications of life on Earth, between 489 and 443 million years ago', which ended in glacially induced End-Ordovician mass extinction.

Whichever growth pattern we adopt, it is not one of uninterrupted increase in diversity with time, for there have been numerous mass extinctions of families, genera and species. The most devastating of these are referred to as 'the big five' – at or close to the end of the Ordovician, Devonian, Permian, Triassic and Cretaceous Periods. The severity of these events may be gauged from some earlier estimates that 96 per cent of all marine species were extinguished at the end of the Permian, around 250 million years ago. Trilobites, graptolites, and certain types of corals, brachiopods, crinoids and foraminifera simply disappeared, while the abundant ammonoids, brachiopods and amphibia stopped just short of extinction. A burst of new life followed in the Mesozoic – corals, resurgent ammonoidea, bivalves, echinoids and reptiles proliferated, only for some like the ammonoids, dinosaurs and certain types of bivalves and marine reptiles to disappear abruptly at the close of the Cretaceous Period, 65 million years ago.

A review of the number of different marine genera (Stanley & Powell, 2003) has reiterated that significant extinctions have characterised not only the End-Permian, but the Ordovician, Silurian, Devonian, and Lower Carboniferous Periods. Each notable extinction was followed by vigorous recovery in the number of new genera. Of particular interest in Stanley and Powell's study was a 50-million-year period of sluggish evolutionary turnover (in both extinction and origination) extending from the Upper Carboniferous to the Early Permian – known as the LPIA or 'Late Palaeozoic Ice Age' – and the migration of some surviving marine genera towards warmer equatorial regions. This manifestation of climatic control is attributed to a global reduction in the amount of atmospheric carbon dioxide, or reduction in the greenhouse-gas effect.

THE CRITICAL IMPORTANCE OF OUR ATMOSPHERE

The question that immediately arises is: what caused the earliest bursts of evolution? In one view (Berkner & Marshall, 1972) it was the oxygen content of our atmosphere. Currently, this is 21 per cent, a level that sustains life as we know it. The paradox here is that in the distant past, a level this high would have destroyed the very amino acids that were essential for the development of evolving organisms. That is, the oxygen-deficient conditions necessary for the evolution of life in the first instance would be unable to support present-day life, and vice versa.

A remarkable paper was presented in 1927, at the twenty-fifth annual meeting of the South African Association for the Advancement of Science, by the Rhodesian geologist, A.M. Macgregor. What made the paper remarkable were his conclusions about the atmosphere in Precambrian times – little different to the views held today, but based purely on fieldwork, a few chemical analyses and simple microscopy, and the meagre library holdings of the fledgling Geological Survey of Southern Rhodesia.

Macgregor concluded that the early atmosphere had been rich in carbon dioxide but highly deficient in oxygen, basing this on his observation that ancient sediments and altered igneous rocks are rich in carbonates, with a high ratio of the more soluble ferrous (less oxidised) compounds to those bearing iron in the more insoluble ferric (more highly oxidised) state.

With regard to the absence of calcareous shells in Precambrian times, Macgregor reasoned that 'there is probably a limiting maximum concentration of carbon dioxide in water above which it is impossible for organisms to form calcareous skeletons. If, moreover, skeletons were formed in such conditions, they would almost certainly be re-dissolved after the death of the animal.' He equated the sudden appearance of fossils with calcareous shells or exoskeletons with a decline in atmospheric carbon dioxide.

From his research into gold-bearing sediments of the Witwatersrand Basin, 2 800 to 3 000 million years in age, Dieter Hallbauer of the South African Chamber of Mines also concluded that 'preservation of pre-depositional geological information . . . necessitates the assumption that an oxygen deficient atmosphere prevailed' (1983: 56). Furthermore, the relatively late appearance of dark red pigmentation in sedimentary rocks of the Waterberg Group in the Limpopo Province of South Africa and Botswana indicates a change to oxidising conditions around 1 800 million years ago (Tankard et al., 1982: 203).

Accepting that the earliest atmosphere was, indeed, effectively devoid of oxygen (Berkner & Marshall, 1972; Sylvester-Bradley, 1972), the scenario offered by Berkner and Marshall is that, under bombardment of the water molecule by ultraviolet radiation from the sun, oxygen (O_2) and hydrogen (H_2) would slowly have been released into the atmosphere, by the dissociation of water (H_2O). The lighter hydrogen would have escaped into outer space, with the heavier oxygen accumulating and effecting the oxidation of the rocks existing on land and beneath the sea. This process of dissociation of the water molecule would, however, have been self-regulating in that oxygen, together with ozone in the atmosphere, would then have begun screening out the ultraviolet radiation that had produced these gases in the first place. Any further increment in oxygen levels above around 0.1 per cent would in this way have been inhibited.

The balance between lethal ultraviolet and the non-lethal radiation of visible light would at this stage have forbidden the evolution of any organisms in direct sunlight, but would have allowed this to proceed at shallow depths in lakes and inland seas, where they would have been protected by the uppermost few metres of water filtering out the destructive radiation. The subsequent development of primitive plants, capable of converting carbon dioxide to oxygen and organic molecules by photosynthesis, would then have been crucial in carrying the oxygen content of the atmosphere to the next critical level of around 1 per cent. The possibility of energy supply to organisms through respiration then appeared – a major step in evolutionary development. Subsequent build-up in atmospheric oxygen became greatly accelerated, and higher ozone levels now allowed life to develop on land as well, by now screened from 'the lethal glare of solar ultraviolet radiation'. Berkner and Marshall equate the burst of marine life in the Cambrian Period with the attainment of the 1 per cent level of oxygen, and the subsequent rapid spreading of both plants and animals on dry land (400 million to 420 million years ago) with attainment of the 10 per cent level.

Subsequent greenhouse effects might then have been controlled by the levels of carbon dioxide in the atmosphere, with the suggestion that the decline in photosynthesis during periods of depletion in this gas might have triggered ice ages, catastrophic reduction in oxygen levels, and the massive extinctions during the Permian, with the disappearance of the huge reptiles at the end of the Cretaceous Period.

At the present time of widespread concern about our environment, the term 'greenhouse gases' is treated as a pair of dirty words. What is forgotten is that without them, Earth would have a temperature in the region of –18 °C and life as we know it would not exist (Nisbet & Sleep, 2001). This illustrates the delicate balance between 'too little' and 'too much' that has raised average temperatures on Earth by some 33 °C, to their present congenial level.

EXTINCTIONS AND SEA-LEVEL CHANGES

A rather different avenue of thought has focused attention on a correlation between extinctions and the changing levels at which the sea has stood during the Phanerozoic – that stretch of time characterised by the record of well-preserved fossils. The survey by Hallam (1992) shows that there has been a slow rhythm or pulse to various phenomena on Earth during the last 600 million years.

These phenomena are remarkably diverse in character, but all appear to show a swing of the pendulum that may be linked to changing sea levels – the prevalence of ice sheets; the deposition on the sea floor of carbonates (limestones), or black shales laid down under oxygen-deficient conditions; the presence of ironstones and phosphate-bearing sediments; the isotopic composition of elements such as carbon, sulphur and strontium in sea water; the carbon dioxide content of the atmosphere; and volcanic activity. Hallam stresses that the controls underlying these changes are not fully understood, but the key appears to be the amount of land mass exposed above sea level, and the effect of this on changes in climate, rates of erosion, weathering, and the transport of debris to the sea.

This slow rhythm or pulse is one that is defined by high levels of the sea during the Lower Phanerozoic (Ordovician and Silurian), then low levels throughout the following Carboniferous, Permian, Triassic and early Jurassic Periods, a return to high levels in the later Jurassic and Cretaceous, and finally a decline once more to present-day levels. On a broad canvas, the time interval between protracted periods of high and low sea level is taken by Hallam (1992) to be roughly 200 million years. High sea levels correlate broadly with more active volcanism, increased volumes occupied by the ocean ridges where new sea floor is created, equable climates, and increases in carbon dioxide in the atmosphere. The rhythm is thus one of alternation between 'icehouse' and 'greenhouse' conditions.

Significantly, Hallam establishes a correlation between sea-level minima and fossil extinctions. The link here appears to have been the area and quality of habitat available in the marine environment. The continental shelves overlain by relatively shallow seas are the places where prolific animal life is concentrated; any reduction in the area of this with falling sea level would have had fatal results.

Theories such as those above attribute extinctions to relatively slow, cumulative effects, and many insist that a massive extinction of plants and animals is actually in progress at the present time. 'There is compelling evidence that the earth is currently in the midst of a human-induced mass extinction . . .' is the view of Miller and Foote (2003) at the Universities of Cincinnati and Chicago. The burning of fossil fuels leading to release of greenhouse gases, encroachment of natural habitats by urbanisation, harvesting of timber in forests, and pollution of the environment are all contributory causes to this postulated extinction.

Not all scientists adopt the role of prophets of doom. Writing in *The Spectator* of 23 February 2002, Matt Ridley reviews the work of the Danish statistician, Bjorn Lomborg, who asserts that claims of species extinction and global warming are either exaggerated or wrong. Ridley suggests, somewhat cynically, that an increasing number of scientists have a vested interest in pessimism, since the issue of global warming has brought them fame, funds and speaking fees: any rejection of their views would be a threat to their livelihoods.

EXTINCTION BY CATASTROPHE

Hypotheses of another type favour catastrophic events as the cause of massive extinctions. The discovery of 'the iridium layer' at the Cretaceous/Tertiary boundary belongs here. Iridium is a metal of the Platinum Group: the name 'iridium layer' derives from the fact that it contains 300 to 500 times the normal iridium levels typical of sedimentary rocks. Often only millimetres in thickness, it has been found in many parts of the world, and dated to 65 million years before present.

Suggestively, space dust also contains about 1 000 times more iridium than the crustal average. The postulate, linked with the name Alvarez (father and son) at Berkeley, California, is that the Earth was struck by a massive meteorite, giving rise to a gigantic dust cloud that encircled the globe and blocked out the sunlight. The most likely site for the impact is considered to be Chicxulub in the Yucatan Peninsula. The resulting decline of photosynthesis is held to have led to the collapse of the food chain and the extinction of, for example, the dinosaurs and flying reptiles (pterosaurs) at the end of the Cretaceous Period.*

True to form in Earth science, a major controversy has flared up around the significance of the iridium layer. The view of Hallam (1989) is that the theory of impact-related wholesale extinction is suspect, noting that further research had suggested that the extinction at the Cretaceous/Tertiary boundary was not as abrupt as first believed. Some have claimed that the extinction was stretched out over several million years rather than the 50-year period postulated. Many reptiles did survive.

The eruption of volcanics of the Deccan in India was the preferred cause cited by some, and the argument was supported by the discovery that airborne particles ejected

* Paul Halpern (1998: 127), a Professor of physics at the Science University, Philadelphia, reminds us of the 'year without a summer' in the aftermath of the eruption of Mt. Tambora in Indonesia in 1815. This massive volcanic catastrophe produced an ash cloud that darkened the skies. Crops in the region were ruined, leading to famine and an eventual death toll of some 88 000 people. The fine ash in the atmosphere drifted over the surface of the globe, reducing the warmth of the sun and lowering the average global temperature. This triggered what Halpern dubs a 'micro ice age' – a summer of bitterly cold temperatures, snowfalls in June as far away as the USA, and accompanying crop failures there too.

during eruption of the Kilauea volcano in the Hawaiian Islands were found to be strongly enriched in iridium. Against this, in turn, minute spherules that might be microtektites (glassy spherules associated with meteorite impact) were found in the clay layer at some places, in addition to mineral grains showing features indicative of shock, and fluffy aggregates of carbon interpreted as soot (from burning forests?). Hallam concluded that 'the extinctions were complex and selective, taking place over an extended period of time, and therefore not plausibly accounted for by a single impact' (1989: 197). A volcanic scenario was therefore the one to receive his vote.

The controversy swings back and forth, but the pendulum has been given a decided push towards the vision of catastrophic impact with an extraterrestrial body by the evidence reviewed by Qixiang (2001: 361–72). The extinction of 90 per cent of shallow bottom-dwelling marine organisms at this time, in addition to all ammonites, 'almost all marine reptiles' and the dinosaurs and other giant reptiles with a body mass of more than 25 kilograms, appears to be matched by massive forest destruction on land. Where 70 to 85 per cent of all pollen preserved in late Cretaceous and early Tertiary strata in Mexico is tree pollen, this proportion has been found to drop to less than 1 per cent in the boundary layer itself, pointing to massive forest destruction at that time.

The scenario presented by Qixiang is that of a meteor with a mass of some 10^{14} to 10^{16} kilograms striking Earth with a velocity in the region of 40 kilometres par second, producing a spreading mushroom cloud at a temperature of 100 000 °C. A train of disastrous reactions would have followed. Plants would have been defoliated on a massive scale. Our protective ozone layer would have been largely destroyed, and the suppression of sunlight imposed by the dust cloud would have inhibited the vital process of photosynthesis by plants for at least several months. Nitrogen oxides produced by the combination of atmospheric nitrogen and oxygen would have yielded acid rain, and acidified the waters of the oceans, killing off the plankton. This destruction or decrease in fertility of the plankton would in turn have reduced the extraction of dissolved carbon dioxide, leading to increased levels of this potent 'greenhouse gas' in both the seas and the atmosphere. This is consistent with oxygen isotope data which does, indeed, indicate a warming of the environment by some 5 °C through the succeeding 40 000 years.

Qixiang (2001) has concluded that the older beliefs are no longer valid – those beliefs that profound changes in the fossil populations preserved in rocks merely reflect gaps in the record, at times when the host sediments were not being deposited, or were eroded away. Charles Darwin's reliance on this approach is challenged, and Qixiang suggests that other iridium anomalies at the boundaries between geological periods originally defined by their fossil populations may in time be shown to be due to similar catastrophic events. As he puts it: 'The discovery of catastrophic events and their great impacts on the evolution of life, therefore, should also be accepted as a revolution in geoscience, because it has liberated the people from uniformitarianism.'

A popular modern compromise holds that the animals and plants that became extinct were already under stress imposed by ecological factors before the impact event merely delivered the *coup de grâce*. This view therefore combines the impact theory of extinction with the idea of well-advanced decline in numbers due to other causes. It might be argued that this demands an extraordinary coincidence in the timing of events, but some have countered this objection by pointing out that the mass extinction was itself an extraordinary event and, as such, a rare conjunction of natural phenomena was not implausible. Whatever the case, the controversy is a modern echo of the conflict between the catastrophist doctrine of Cuvier, and a slow progress of events as in the approach by Lamarck and Darwin.

20
BLACK SMOKERS AND WHITE SMOKERS

'Most biologists now believe that life originated by a process of slow development from non-living chemical systems; today few areas of science offer more challenge and excitement . . .'

A. Lee McAlester (1968: 4)

This topic has nothing to do with the effects of nicotine, nor the habits of different groups of addicts. 'Smokers' in the geological lexicon are emanations of hot water or brines from the rocks of the sea floor, discharged into the overlying waters of the oceans. Where they discharge from point sources, they are clear, hot solutions at temperatures of 150 °C to 400 °C that rise and then mix turbulently with cold sea water, precipitating the dissolved metals they carry as black or white 'clouds' or plumes of sulphides, hydroxides and sulphates. They may build mineralised chimneys of sulphides of iron and copper above the point of discharge, and the heated waters then rise hundreds of metres to disperse into the sea water above. A more unfocused form of discharge also emerges as seepages from networks of fractures on the ocean floor, generally at lower temperatures and with lower contents of dissolved mineral matter.

These systems are attributed to circulation, by convection, of sea water through lavas of the sea floor, mainly at the boundaries between the tectonic plates into which the Earth's crust is broken. Cold water is drawn down into the fractured and pillowed lavas, and heated. Metals present in the sea-floor rocks are taken into solution here during the process of circulation, and the heated water vented again from the ocean floor as it rises through the system.

This is a more than trivial process within the spectrum of geological phenomena. Writing in *EOS, Transactions of the American Geophysical Union*, Rona and co-authors (2002) estimated that an amount of water equivalent to the entire mass of the ocean may be circulated through plate boundaries within a period of 10 million years, and claim that the processes control '. . . budgets of major and trace elements in sea water . . . release carbon dioxide from magmatic systems through the ocean into the atmosphere in quantities that may contribute to climate change; [and] concentrate metallic mineral deposits . . .'

Emanations of a different type – of fluids and gases, notably methane – which should not be confused with those described above, are now receiving much attention. Seepages of water, brine, oil and gas emanating from accumulations of gas hydrate (ice-like solids charged with gases) in sea-floor sediments in the Gulf of Mexico and off the west coast of the Iberian Peninsula have produced mounds and mud volcanoes (Somoza et al., 2002). These organic fluids appear in large measure to be of biogenic origin. Surveys using geophysical techniques and underwater photography have revealed mud volcanoes as large as 2.5 kilometres in diameter, rising up 300 metres from the sea floor. Estimates of the total amount of methane gas stored in this way in the sea floor indicate a vast potential energy resource – somewhere in the range 3 x 10^{15} cubic metres to seven times that amount (Milkov et al., 2003).

LIFE ORIGINATING IN A VOLCANIC SETTING NEAR BARBERTON?

The topic of smokers might, at first glance, appear to offer little that would excite interest in the context of southern African geology, with its abundance of deep-seated granites and gneisses, ancient layered and unlayered igneous complexes, sediments of terrestrial origin, and huge volumes of Karoo basalts poured out on the sands of a great desert. If we turn, however, to the relics of our most ancient rocks, such as those around Barberton, the topic leads us straight back into speculation about the possible origin of life on our planet in the remote past.

Within the Barberton region, there are rocks known as the Onverwacht and Fig Tree Groups that have revealed signs of primitive life forms (bacteria and algae) considerably older than 3 250 million years. In their account of some of the features exhibited by these very old rocks, De Wit et al. (1982) describe carbonaceous layers that could represent mats of ancient organic matter, and domical structures (stromatolites) attributed to algal growth, all of which are suggestive of a biogenic origin.

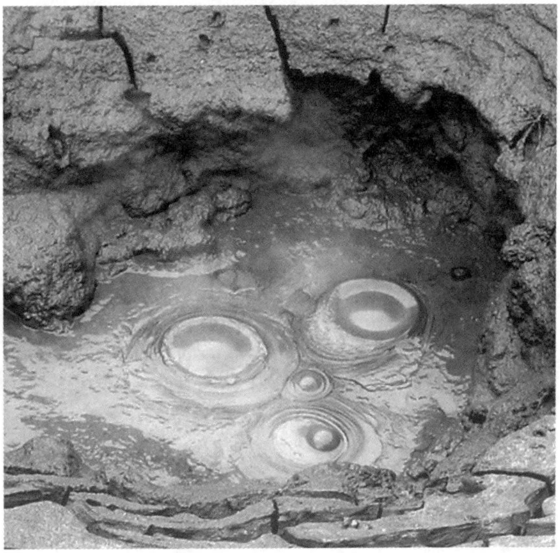

Figure 35: Rings created by gas bubbles rising to the surface of pools of black, sticky mud at Hell's Gate, in the volcanically active area of Rotorua, New Zealand. Structures resembling these have been compared by De Wit et al. (1982) to those found in association with material believed to be fossil mats, of organic origin, in the ancient rocks of the Barberton region.

Associated with these biogenic relics are enigmatic, concentrically ringed structures in the ancient sediments, resembling piles of nested saucers of progressively smaller diameters, bearing a remarkable resemblance to the patterns that are produced at the present time within mud pools in geothermal terrains, such as, for example, those in the volcanic region around Rotorua in North Island, New Zealand (see Figure 35). Here, bubbles of heated gas rise through the sticky muds to produce a ring as they break surface, with each ring then expanding in size as the next bubble follows. Thus, the association is made here between heated gases rising in volcanically active regions, and fossil life forms in the ancient Barberton rocks.

In the same area there are what De Wit and colleagues refer to as 'ironstone pods', some being roughly circular with diameters up to 50 metres, and others as elongated shapes up to 100 to 200 metres in length. These are composed largely of hydrated iron oxides (the mineral goethite), manganese oxides and clay minerals, and were interpreted as 'buried hydrothermal chimneys or channelways to hydrothermal vents'. It was mooted that sites of hydrothermal activity, such as these were claimed to have been, may have been important environments for the development of prolific biological activity in the Archaean, much as they are found to be around deep ocean springs today.

This thesis has drawn support from later studies of organic matter that have been found to be associated with black smokers of a variety of ages. In one such study, De Ronde and Ebbesen (1996) found essentially the same organic compounds at different sites, irrespective of their ages, to conclude that 'present-day sea-floor hydrothermal systems might indeed be representative of the first habitat of life'. The implication is that 'highly evolved' micro-organisms were already in existence more than 3 250 million years ago, and perhaps as early as 3 800 million years ago, in the environment within which these particular rocks of Barberton were deposited. This led to a further assertion, based on a study of samples of the Fig Tree shales, that 'a photosynthetic, pelagic, marine biota must also have been vigorous at this time'.

A study by six co-authors (Westall et al., 2001) has reviewed earlier evidence on which the claim to have discovered fossil micro-organisms in these ancient sediments was based. Where some earlier researchers had been sceptical about the identification of the minute structures as fossilised living forms, Westall et al. have taken a more positive line. Using scanning electron microscopy and isotope analysis of the traces of carbon in these rocks, and a comparison of the morphological features with those displayed by known bacteria, they have asserted that structures 'probably representing fossil bacteria or phenomena associated with bacteria' exist here. They describe minute spherical particles 1/1 000 millimetres (1 μm) in diameter, occurring singly or as clusters and strings, and rod- or sausage-like forms up to 3.8 μm in length, occurring in colonies. Their conclusion is that the data from the Swaziland rocks indicate 'quite a wide diversity of probable and possible microfossils during the Early Archaean . . .'

A REBUTTAL

A reappraisal of the origin of the ironstone pods in the Barberton rocks has suggested applying the brakes to the ideas mooted above. Lowe and Byerly (2003), writing in the journal *Geology*, have rejected the interpretation of the ironstone pods as ancient features related to smokers, and attribute them to spring activity in very much later times, perhaps during wetter periods within the Pleistocene.

They concede that while hydrothermal vents may well have been important for biological evolution in times past, the Barberton pods do not represent an ancient example. They protest that the mineral species, goethite, of which the pods are largely constructed, can by microscopic studies be proved to have been one of the first species to have been formed at the site, yet its presence here is simply incompatible with its inherent instability at temperatures above 80 to 100 °C. The surrounding Archaean rocks can be shown to have been heated to temperatures of at least 200 to 250 °C during Archaean times. Furthermore, goethite is unknown in rocks older than the Ordovician.

Fieldwork carried out at the site would also seem to support a very young age for the pods. The cavernous nature of the deposits has allowed the growth of stalactite- and stalagmite-like structures formed within cavities by water-drip, and the present vertical orientation they exhibit (dictated by gravity) is in conflict with the tilted attitudes of the Archaean strata. This led Lowe and Byerly to the conclusion that the pods are superficial cappings of no great age that tell us nothing about conditions during the Archaean. We can only wait for the riposte that must surely follow this thrust.

Plate 1: Photograph of varved sediments of the glacial Dwyka Group in Natal, illustrating fine-scale alternations of lighter and darker layers, each a few millimetres thick, representing seasonal variations (spectacles placed as scale object).

Plate 2: Fossil leaves of the Permian plant, *Glossopteris*, an important element in the development of the concept of continental drifting. (Photograph courtesy of Dr Rose Prevec.)

Plate 3: The flat-lying strata of the Karoo Sequence in the Sani Pass area of KwaZulu-Natal, typical of the entire region extending southwards into the Eastern Cape province. Stow had in 1870 interpreted this present-day landscape as the result of glaciation in fairly recent times – a proposition that is fallacious.

Plate 4: George Stow (1870) had attributed many common erosional features to glaciation in the relatively recent past. The characteristically dome-like outcrops of Karoo dolerite intrusions, produced by exfoliation of curved slabs from the larger outcrops, he had interpreted as *roches moutonnées* – hummocky outcrops shaped by ice abrasion. This example in the Mountain Zebra Park, near Cradock, is typical of the outcrops that would have misled him.

Plate 5: The enormous power of the forces that deformed the rocks of the Cape Supergroup during plate movement is emphasised by the buckling and tilting of even the strongest strata. These strata, now twisted like *pasta fettuccine*, were originally laid down as horizontal layers. The locality is Waaipoort, Steytlerville district.

Plate 6: The drifting of the continents, as interpreted in the Plate Tectonic Model, was responsible for the rising of the majestic mountain chains of the Cape Fold Belt in the southern Cape province. Sedimentary rocks that were laid down in the Agulhas Sea that existed at that time, between 500 and 330 million years ago, were later squeezed, intensely crumpled and uplifted by enormous forces directed from the south during plate movements. The photograph, taken in Meiringspoort, south-east of Prince Albert, shows the contortions laid bare in the gorge carved by the river running through it.

Plate 7: The Dwyka Tillite as it appears in its southern outcrops in South Africa. The included pebbles and fragments of other rock types trapped in it (white) are conspicuous. In earlier days, some saw it as a volcanic or intrusive rock while others attributed it to bizarre modes of sedimentary deposition. Only in 1868 did P.C. Sutherland correctly recognise it as a product of glaciation formed before the fragmentation of the great southern continent, Gondwanaland.

Plate 8: Convincing evidence pointing to a glacial origin for the Dwyka Tillite – grooves and scratches gouged into the underlying Ventersdorp lava in the northern regions by advancing continental ice sheets armoured with rock fragments. The locality is Nooitgedacht in the Kimberley district.

Plate 9: Table Mountain from the cable car, showing the Table Mountain Group, dominated by hard quartzites, with which the rocks of the Witwatersrand gold mines north of the Vaal River were mistakenly correlated in the 1890s on the basis of general lithological similarity.

Plate 10: The extensive body of water that Schwarz sought to restore to the Etosha Pan, by diverting major rivers, is today represented by only scattered waterholes at which animals drink.

Plate 11: Aerial view of the Tswaing Crater (photograph by courtesy of Professors W.U. Reimold and R.L. Gibson, University of the Witwatersrand).

Plate 12: Fossil trackway (footprints) of a mammal-like Dicynodont reptile that tramped through mud of Late Permian age, 253 million years ago. The pencil below the centre of the picture gives the scale.(Photograph by courtesy of Dr W.J. de Klerk, Albany Museum, on the farm 'Sante (Asante) Sana', east of Graaff-Reinet.)

Plate 13: Despite the striking resemblance to a plant with stems bearing leaves, this is not a fossil, but a dendrite formed by groundwater depositing iron and manganese oxides within a fracture in banded ironstone, laid down more than 2 billion years before the first land plants evolved. What is unusual is the manner in which the dendrite stands out in bas-relief, rather than as a mere stain on the surface. Original specimen collected in the Kuruman district by Dr. Stefane Polteau of the Rhodes University Geology Department.

Plate 14: Chalcopyrite, the common copper sulphide mineral found below the zone of weathering, showing characteristic golden-yellow colour.

Plate 15: Native copper (reddish) intergrown with characteristically green secondary copper minerals that form at shallow depths above copper deposits.

Plate 16: Tiger's Eye, a semi-precious gemstone formed by silicification (replacement by SiO_2 or silica) of fibrous asbestos in deposits close to surface. Although a stone of great beauty, its value is degraded by the abundance in which it is found.

Plate 17: The beauty of natural crystals, whose shape and growth are dictated by simple mathematical laws. This is quartz, or crystalline silica, otherwise known as 'rock crystal', or at one time as the 'Cornish diamond'.

Plate 18: Early copper workings (now abandoned) at Tweefontein in Namaqualand. The size of the excavation originally occupied by the orebody is indicated by the bald hill in the background, with a large group of students barely visible below the summit.

Plate 19: The Kimberley 'Big Hole' as it is now. The upper half of the photograph shows the widening of the excavation where the more easily decomposed Karoo Supergroup strata have collapsed; the lower half shows the deeper, older Ventersdorp Supergroup rocks through which the pipe blasted its way. Williams (1932) gave the surface area of the excavation as 38.2 acres (15.5 ha) at that time, and the level of the water as 933 feet (285 m) from surface. (Photographed by Lianne Butler.)

Plate 20: A near-vertical 'fissure' or dyke of kimberlite, less than a metre in width (lighter coloured rock in centre of photograph), intruded here into horizontally bedded dolomitic limestones of the Transvaal Supergroup at Bellsbank, Kimberley district.

Plate 21: Rock crusher used by 'the Ancient People' of Zimbabwe in prehistoric times, in the Gwanda district. The great boulder was rocked over pieces of gold-bearing quartz ore by inserting a log into the groove chiselled into its surface, and using the small pillar on the left as a fulcrum.

Plate 22: Close-up view of the famous Witwatersrand banket – the gold-bearing conglomerate that is the major source of South Africa's gold. Rounded pebbles a few centimetres in diameter are cemented together in a finer-grained matrix, together with abundant brass-coloured pyrite (iron sulphide, FeS_2).

Plate 23: Spectacularly rich gold ore of the type that may be found in the Basement rocks of Zimbabwe or the Barberton region.

Plate 24: The coarsely crystalline, platinum-mineral-bearing Merensky Reef resting upon feldspar-rich anorthosite (feldspar rock) deep underground in Union Section Mine, Northwestern Bushveld Complex. The 'reef' is normally a metre or less in thickness, but together with the UG2 Chromitite Layer at a deeper level, currently supplies 60 per cent of world needs, and holds close on 88 per cent of its reserves of the PGMs, according to statistics for 2003 compiled by the Department of Mines and Energy. The scale object near the bottom-right corner is a pencil pointing into the Reef.

Plate 25: The UG2 Chromitite Layer (which appears black) exposed in opencast working, western Bushveld Complex. The potential ore reserve of this layer is now known to exceed that of the Merensky Reef by a factor of 1.5.

PART FOUR

Rocks, Minerals and our Unmatched Ore Deposits

21 INTRODUCTION: OUR EXTRAORDINARY MINERAL WEALTH

The rise of what is now the Republic of South Africa, from what was once an impoverished cluster of colonies and virtually bankrupt independent republics in the mid-nineteenth century, into the most industrially advanced state in Africa, is to be attributed in large measure to the discoveries of its extraordinary mineral wealth.

First came the finding of diamonds in 1867, a veritable cornucopia that yielded 137.9 million carats (around 31 short tons) of the precious stone up to the year 1913 alone. The discovery of the world's greatest gold field – the Witwatersrand Basin – followed in 1886, and by 1984 this had realised a massive recovery of just short of 40 million kilograms of the yellow metal. In time this field was destined to have produced some 40 per cent of all the gold ever mined in the world. With the coming of the 'atomic age' it emerged that the ancient gold-bearing Witwatersrand conglomerates also ranked fourth in the world as a reserve of uranium.

The discovery of economic deposits of platinum-group metals in the Merensky Reef in the Bushveld Complex followed in 1924, and with it the realisation that South Africa holds the greatest resource of these metals on Earth. It also emerged that the Complex hosts more than half of the world's resources of chromium, 60 per cent of its vanadium, and a huge untapped resource of titanium.

The sales value of all minerals produced had reached R4 134 million for the year 1975, which burgeoned to R70 800 million for the year 1998, while an anticipated average growth rate of 4.3 per cent per annum was forecast up to the year 2004 (Minerals Bureau, 2000). The actual sales value realised outstripped this forecast to the extent that total sales in 2002 had already reached R134 765 million. A decline in sales revenue to R117 750 million in 2003 is a reflection of the strengthening of the rand from R10.52 to the US dollar to R7.56 over that period. Without this, the sales revenue would have exceeded R160 billion.

Inevitably, there is a background to the development of our mineral deposits that is rich in controversy, contains conflicting ideas on their origin and value, and generated both good and bad predictions of future prospects. The discovery of diamonds in South Africa was at first stridently denounced as fraudulent, or attributed to the regurgitation of the stones by wandering ostriches. Some initial forecasts in the earliest days had pessimistically claimed that the Witwatersrand gold deposits would peter out at a depth of five metres, while another in 1890 predicted the collapse of the entire gold-mining industry. Where platinum was at one time regarded as a worthless metal, the story of its rise to its present position of eminence reads like a fairy tale.

In Earth science the physics of our planet, the hatching of new theories, the world of fossils, and the origin of life do not hold the monopoly on controversy, dogmatism or fascination.

Alloyed with tin, copper yields bronze, a metal used by humans since ancient times. It was in use in the Aegean region in the fourth millennium BC, and in Western Europe from about 1500 BC (Bazin, 1958: 20). Masons and skilled metallurgists supplied by King Hiram of Tyre in Phoenicia were employed in the building of the Temple of Solomon on Mount Moriah, Israel, in 962 BC. Such was the skill of these craftsmen that they cast hollow bronze pillars with a diameter of 1.9 metres, capped by extravagantly ornamented capitals, reaching an overall height of over 12 metres (Andrews, 1999: 40–47). Andrews tells us that this bronze was an alloy of copper, tin and arsenic, and that its value was about one quarter of that of gold. The use of bronze for weapons and statuary in Roman times needs no elaboration here. In South Africa, bronze ingots have been recovered at Rooiberg

(Recknagel, 1908) where tin was mined in ancient times, while copper appears to have come mainly from Messina (Musina) in the Limpopo Province (Söhnge, 1946) and Phalaborwa (Hall, 1912), with mining probably dating back to the third century AD (Miller, 1995). Plates 14 and 15 illustrate two common ores of copper that would have been used in ancient times.

NATURAL MATERIALS USED BY HUMANS IN EARLIER TIMES

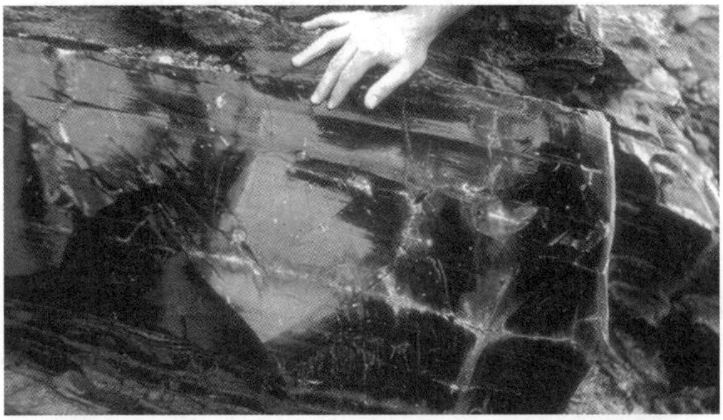

Figure 36: Natural black glass, obsidian, a rapidly chilled, viscous lava rich in SiO_2. The lines passing through this mass are due to layers of small crystals stretched out while the magma was still capable of flowing. Pieces broken off yield fragments with razor-sharp edges that ancient humans used for making knives, arrowheads and scrapers. The locality is Newberry Crater, Oregon, USA.

Figures 37 and 38: The ruins of Sacsayhuaman in Peru, a fortress built by the Incas in the late fourteenth or early fifteenth century. What is remarkable is that the massive stone blocks are not rectilinear in shape, which would have made the fitting together of them less of a three-dimensional problem. Some blocks are triangular, others quite irregular in shape, yet they are fitted together so snugly that a knife blade cannot be inserted between them (right-hand photograph). How this was achieved with primitive technology, or how the great blocks were then lifted into place, have never been convincingly explained.

22
Tablets of Stone

'And he hewed two tables of stone like unto the first; and Moses rose up early in the morning, and went up unto mount Sinai . . .'

Exodus 34: 4

'And he wrote upon the tables the words of the covenant, the ten commandments . . . Moses came down from mount Sinai with the two tables of testimony in Moses' hand.'

Exodus 34: 28–29

'All the commandments which I command thee this day shall ye observe to do, that ye may live, and multiply . . .'

Deuteronomy 8: 1

In the 1940s, before being elected to a Chair of geology at Columbia University in New York, one of the leading petrologists of southern Africa was on the staff of the Bechuanaland (subsequently Botswana) Geological Survey. In conversation with him there one afternoon, he offered a novel interpretation of the previous Old Testament passages that should be interpreted as a reaching for a scientific interpretation of an ancient narrative.

It was this petrologist's contention that the tablets of stone were in reality specimens of the rock type known as graphic granite, which Moses had selected knowing that his flock were unable to read written text, but in dire need of a rigid code of social conduct. Such a code, as clearly set out in Exodus 20 and enlarged upon in *The Book of the Covenant*, covers diverse topics such as proper conduct, employment of servants, and penalties for theft, assault, murder and adultery – a code for decent and upright living.

Graphic granite is a rock that shows a highly distinctive intergrowth of the silica mineral, quartz (SiO_2) and potash feldspar (commonly white or pink), in which the shapes of the quartz crystals on a broken or polished surface show, at first glance, some resemblance to ancient cuneiform or Hebrew script. To the untutored, this could pass as a written record of instructions that had been delivered verbally. Ancient Basement rocks, including granites, occur on both sides of the Red Sea and in the Sinai Peninsula, thus the proposition is not ruled out on geological grounds. Hence, this could conceivably be regarded as a stratagem whereby Moses communicated his wisdom to his people, and gave them a sound code of conduct to follow.

Figure 39: Drawing of a surface of the natural rock type known as graphic granite, displaying the remarkable texture suggestive of some ancient text. The 'letters' drawn in black are the mineral quartz, and the unshaded areas are the light-coloured feldspar with which the quartz is intergrown.

23
Gemstones: Supposed magical properties

Whether you believe in fairies, leprechauns and trolls, or not, Sue Eales's (1979) light-hearted piece on Medieval English literature, drawing on the compilation by Evans and Serjeantson (1960), reveals the ancient mysticism that enveloped the mineral kingdom in bygone days.

> Would you believe that . . . emeralds could cure gout and lechery? . . . or that agates could promote clear speech? . . . or that sapphires could cure a bleeding nose, or kill spiders?
>
> Medieval man took his lapidary [the polishing and use of precious stones] very seriously, and in literature precious stones were popular for their beauty, their symbolic significance, and their curious supposed virtues. Various lists of stones were compiled, giving a description, locality and the medicinal properties for each. Alabaster, for example, was described as an ashen-white stone from Egypt, which when powdered, dissolved in vinegar and drunk would cure sores on the feet and knees. By drinking water in which beryls had been immersed, a man could cure hiccuping, liver disorders and eye infections. Amethysts, when worn, were supposed to offer protection from wild beasts, witches and imprisonment, and prevent his horse from collapsing or falling prey to worms. They also enhanced his singing. If he were imprisoned, and could touch his bonds with a sapphire, he would be magically released. He could be made invisible by hiding an agate in a marigold and clenching it in his fist.

A chrysolite could act as a passport into strange countries for the virtuous man, and a jacynth would ensure his safe passage. A topaz would safeguard his chastity, and an emerald increase his wealth, while a ruby comforted his heart and body. An onyx enabled him to converse with the dead, and a jasper prevented wild fantasies. If all this were not enough to satisfy him, an onyx could also facilitate his spitting!

Medieval ideas about the formation of various stones appear rather fanciful to us now: it was believed, for example, that pearls were formed as a result of the dew of heaven being received by shellfish at certain times of the year. Rubies were believed to be descended from the holy flames of Paradise. Coral was thought to be a stone made from the foam of the sea, which was borne by the wind into the air, where it congealed and fell to the Earth in sticks. These were then used as a safeguard against thunderstorms. The colour of a topaz was supposed to wax and wane with the moon. Because of this, a topaz would help in the cure of lunacy. If, for some reason, one had to place one's hand in boiling water, dropping a topaz into the water would make this safe, although the stone would consequently lose its colour.

Diamonds were believed to be of two sorts: the brown or violet ones from India being males, and the whiter ones from Arabia being females. Both were thought to be effective for promoting the correct formation of the bones, and for preventing grievous dreams and temptations.

There was a stone called Magnet, the property of which was to cause iron to spring up against it. This stone was useful for testing the faithfulness of a wife: if it was placed under her pillow, she would kiss it in her sleep if she were chaste; if she had been a 'spouse-breker', she would fall out of bed.

Perhaps the last word on lapidary science must have been the view held by Chaucer in his 'Tale of Melibee': 'What is bettre than gold? Jaspre. What is better than jaspre? Wisedoom. And what is bettre than wisedoom? Womman. And what is bettre than a good woman? Nothyng!'

24
MINERALS: WHAT ARE THEY?

'Avicanna (980–1037 AD) the Arabian translator of Aristotle, grouped minerals as "stones, sulphur minerals, metals and salts".'

J.F. Healy (1986: 114)

'... there are many arts and sciences of which a miner should not be ignorant. First there is Philosophy, that he may discern the origin, cause, and nature of subterranean things.'

Georgius Agricola (1494–1555), mining engineer and metallurgist

'Increasing confusion in nomenclature has ... attended the recent progress of the science ...'

James D. Dana (1868), Preface to the fifth edition of his System of Mineralogy *(E.S. Dana, 1904)*

The term 'mineral' has been so widely and carelessly abused that popular usage has come to pay scant attention to its formal definition as a naturally occurring solid with a crystalline structure that determines its form or shape, its properties and its limits of chemical composition. Extravagant claims on the boxes of breakfast cereals or labels for toothpastes, health tonics, hair shampoos and 'mineral waters' that 'these will supply all the minerals that your body needs' are merely slogans scribbled by commercial advertisers careless of the difference between a mineral and a chemical species or medicinal compound.

FROM ARISTOTLE TO AGRICOLA

Unlike geology, which entered the sphere of science relatively late, mineralogy is one of the older sciences, traceable back to Aristotle's pupil, Theophrastus (368–284 BC). In its beginnings no clear distinction was drawn between minerals, rocks and fossils – fossils were regarded by some as no more than odd shapes adopted by minerals. It is to Pliny the Elder (23–79 AD) that we owe an overview of mineralogy in its infancy during Roman times. Here was a compulsive but uncritical collector of 'facts' in an astonishing range of fields (medicine, botany, pharmacy, zoology, astronomy, chemistry and mineralogy) before his insatiable curiosity drew him into entrapment and death in the lethal hail of volcanic ashes that smothered the ancient town of Pompeii.

Born into a well-connected family, and drawing on his experiences as an army officer, prefect of the Roman fleet, and public administrator, Pliny had poured into his *Natural History* some 20 000 'factual' entries reputedly drawn from 2 000 books by 100 different authors, but the judgement of Healy (1986) is that the Romans made little real contribution of any scientific value. Perhaps we judge Pliny too harshly – surprising maturity is to be found in his treatment of some properties of minerals: their colour, cleavage, hardness, tenacity, magnetism, refraction of light, and pyroelectric properties. There are glimmerings of the concept of mathematical crystal symmetry. But there is little value beyond mere entertainment in the nonsense he wrote about the genesis of rock crystal (quartz or SiO_2) which, he asserted, 'is hardened by excessively intense freezing . . . it is a kind of ice . . . [and] the inevitable conclusion is that rock-crystal is formed as moisture from the sky falling as pure snow' (Healy, 1986: 114). The arcane belief that some minerals and rocks – diamonds, galena and limestone – could regenerate themselves after removal from the ground seems to have been accepted without question by him.

The 'Historical Note on the Theory of Ore Deposits', compiled by Herbert C. Hoover and Lou H. Hoover (1912: 43–53) in their translation from the Latin of *De Re Metallica* by Georgius Agricola (1494–1555), reveals the muddle still prevalent during the time of the alchemists. Before Agricola's treatise, the earliest Renaissance work devoted to ore deposits and minerals was a German text, *Bergbüchlin* (published around 1520) which offered the extraordinary philosophy that 'every metallic ore receives a special influence from its own particular planet . . . thus gold is of the Sun or its influence, silver of the Moon, tin of Jupiter, copper of Venus, iron of Mars, lead of Saturn, and quicksilver of Mercury'. Despite these absurdities, we may still perceive the beginnings of understanding in the conclusion that 'vapours, called mineral exhalations, are drawn up from the depths of the earth . . . and the rising fumes pass into the veins and stringers . . .' to deposit minerals there.

Agricola was a prolific writer, mining engineer, and undoubted genius of his time, on whom Hoover and Hoover confer the ultimate accolade of having made 'a much greater step from what had gone before than that of almost any single observer since'. In the 35 years that followed publication of the *Bergbüchlin*, Agricola developed the theory that openings in the Earth (*canales*) were formed through erosion by 'subterranean waters' where there was 'subterranean heat'. The fillings deposited within the *canales* he separated into 'earth', 'solidified juices' (in a broad sense, salts), 'stone', 'metals' and 'compounds', all deposited from water and 'juices'.

Agricola's thinking was rooted in the basic conviction that the properties of all materials reflected some combination of the four fundamental 'elements' – earth, water, fire and air – each endowed with its own distinctive properties of heat or coldness, dampness or dryness, heaviness or lightness, and rigidity or fluidity. As the translators repeatedly emphasise, his text is far from clear, but his reasoning may be perceived in his argument that all metals must contain water, since they flow like water when melted. Minerals contained either 'solidified juice' and 'stone', or 'metals' and 'stone' that were grown together by the action of 'coldness'. Transparent ruby silver, for example, was of 'silver juice' and juice that 'formed the substance of transparent stone'.

THE FOG BEGINS TO LIFT

The way in which the fog cleared in succeeding years is a long and complex story. It was the Dane, Nicolas Steno (1638–86), revered across Europe as both an anatomist and an observer of rocks before entering the Church as a Bishop, who took the first tentative steps in a study of the geometry of natural crystals – the field of crystallography. Then came a Professor at the University of Paris, the Abbe René Haüy (1743–1822), who started the mathematicians, physicists and chemists on a quest that still goes on today. He theorised that if it were possible to break crystals repeatedly into progressively smaller fragments along their natural planes of cleavage, there would eventually be revealed the *moléculé intégrante*, or smallest possible unit. The way in which these ultimate units pack together to fill space would be the feature that determines the overall crystal shape. This was a massive step forward, for it introduced a rational way of explaining the way crystals grow, and why angles measured on them obey simple mathematical laws.

By the tally of Ford (1918) of Yale University, no more than 200 different minerals were recorded by the year 1800, a year after the formation of The American Mineralogical Society. (With an eye to practical issues, the inauguration of this prestigious body had been accompanied by requests in the press for information on 'stones suitable for gun flints', sulphur, saltpetre and deposits of lead – perhaps an early hint of the American 'gun culture' of later years?) In 1912 came the great breakthrough when Professor von Laue at the University of Munich demonstrated that planes of atoms within minerals will reflect X-rays. Physics then became fundamental to mineralogy, for here was a technique for unravelling mineral structures at the atomic scale, and William Bragg in Britain was to rise to knighthood for his pioneering work in this field.

Well over 3 000 naturally occurring minerals are now known, and the scientific journals offer fresh discoveries all the time. Each species has its own characteristics – optical and other physical properties (hardness, colour, specific gravity, refractive index, etc.), spacing between planes of atoms, chemical composition, or characteristic environments in which they grow. Classification can become something of a nightmare, and here the practising field geologist may be forgiven for adopting the philosophy of the palaeontologist who wrote that 'oysters are meant to be eaten, not classified'!

A GOOD REASON FOR GETTING IT RIGHT

From what has been presented above, the definition of 'a mineral' should be clear, and without ambiguity, but it becomes clouded when governments' lawyers put their oar in. The South African Minerals Act of 1991 defines a mineral as 'any substance, whether solid, liquid or gaseous form, occurring naturally in or on the earth, in or under water or in tailings and having been formed by or subjected to a geological process, excluding water, but including sand, stone, rock, gravel and clay, as well as soil, other than topsoil . . .' (Unattributed, 1993).

Realistically, a definition as broad as this would not exclude any rock or fossil, or for that matter, even a potato, from being regarded as a mineral. The article referred to here goes on, for example, to quote a submission that was made in a court case: '. . . I confirm that the said substance is geologically known as andesite or andesitic lava of the Ventersdorp System . . . In my opinion the said andesite is a mineral in terms of the definition . . .' We might question whether a definition as non-scientific and all-embracing as that above represents any substantial advance on the scheme of Agricola in the sixteenth century, or even what was taught by Aristotle or Pliny. The problem of defining a mineral would in some spheres appear to have come full circle, ending back where it began more than 2 000 years ago.

25
ASBESTOS: VILLAIN OR SCAPEGOAT?

'The critics say further that mining is a perilous occupation to pursue, because the miners are sometime killed by the pestilential air they breathe; sometimes their lungs rot away . . .'

Georgius Agricola (1556)

'. . . there should be a growing demand for short fibre of all varieties which will allow mining to be carried out on a more economic basis. It is expected that "Transvaal Blue" will secure a permanent footing in the world's markets.'

Geological Survey of South Africa (1940: 192)

'On January 29, 1986, the United States Environmental Protection Agency proposed banning five asbestos-containing products . . . and phasing out all remaining asbestos uses in the US over the next ten years.'

Asbestos Institute (1986: 5), Montreal

'The production of amosite in South Africa dropped to zero in 1993, as did that of crocidolite in 1998. Production of chrysotile asbestos dropped by 76 per cent during the period 1989–98.'

Minerals Bureau (2000: 196), South Africa

Predictions of a rosy future for the asbestos industry that had appeared to be unassailably justified in the middle of the twentieth century, and fully meriting recognition as a 'good call', turned to dust because of factors that had little to do with the geology of the natural deposits, the efficiency of recovery, or reserves of the material. The previous four quotes tell the story. The collapse of the asbestos mining industry in South Africa came about with the accumulation of incontrovertible evidence that proved a link between respiratory disease and contact with the material. The bans eventually proposed by the US Environmental Protection Agency (EPA), and by the Health and Safety Executive (HSE) in Britain, were merely the *coup de grâce* in a campaign that had raised a hornets' nest in all asbestos-producing countries, and amongst the industries dependent on supplies of the raw material.

The mining of asbestos began in South Africa in 1893, but several decades were to pass before production became significant. The sales value of the product in 1911 was a mere £20 000, a quite trivial sum compared with, for example, the £8.75 million yielded in the same year by diamond mining. By 1940, the sales value had increased 25-fold to reach close on half-a-million pounds; by 1960 it was close on £11 million; and in 1975 it was R91.5 million. A peak in sales value was reached in 1990, when sales amounted to R195 million, and the future appeared secure. All predictions, however, were worthless in the light of what followed.

The problem with the asbestos minerals is that the fibres are so fine that they may be carried in suspension in moving air, and the more the material is handled, milled or processed, the more the tiny fibres are released. On inhalation, the fibres lodge in lung tissue. In the early days of mining and then treatment of the product in factories, the malady was seen as no different to that contracted by stoneworkers, coal and gold miners, or workers in granaries, smelters, potteries or textile mills, and engravers of cut glass. It was regarded as a form of phthisis.

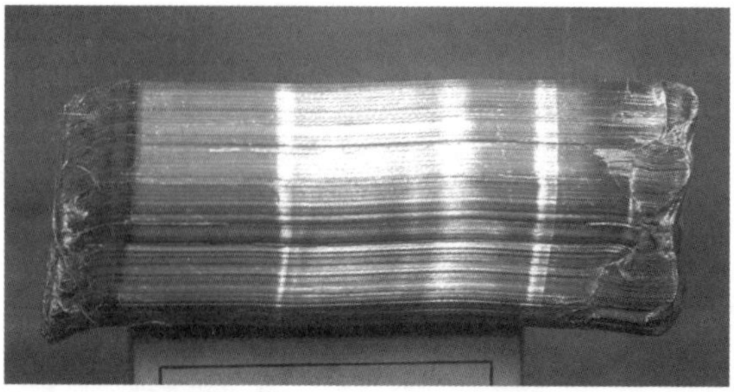

Figure 40: The silky-fibred asbestos mineral, crocidolite, which was banned in most countries of the world because of the link between the mining, treatment and use of it, and respiratory disease, notwithstanding its superior properties in industrial products. When silicified, this becomes the semi-precious gemstone, Tiger's Eye (see Plate 16). (Author's photograph reproduced here with permission of S.A. Council for Geoscience.)

THE 'WONDER MINERALS'

The term 'asbestos' is used, broadly, to include several varieties of different mineral species which occur as hair-like crystals (fibres) so slender but strong that they are highly flexible and can be spun into thread, woven into textiles, or compressed into mats of various sorts. Their ability to withstand high temperatures without undergoing change renders them eminently suitable for the manufacture of fire-resistant curtains and clothing, and of insulating media in electrical installations. Mixed into cement products (asbestos cement), plastics (vinyl floor tiles) and paints, they impart high tensile strength. They are poor conductors of heat, and are also unaffected by sea water, acids and alkalis, and were therefore at one time regarded as a group of 'wonder minerals' with an endless range of uses including the manufacture of carpets, ladies' evening dresses, protective clothing, gas masks, dental fillings, cigarette filters, water filters, etc. It is even said that mattresses stuffed with asbestos were at one time put on the market for the convenience and reassurance of those addicts who habitually smoked in bed! It was a scene guaranteed to reduce any twenty-first-century Environmental Control Officer to apoplexy.

Two quite distinct mineral groups, with different chemical composition and physical properties, fall under the sack-name of 'asbestos'. Those belonging to what mineralogists term the Amphibole mineral group include ash-grey amosite (the name derived from **A**sbestos **M**ines **o**f **S**outh Africa), montasite, and the blue asbestos, crocidolite (Transvaal Blue and Cape Blue). These occur in ancient banded ironstones of the Transvaal Supergroup, more than 2 400 million years old, within a strip of terrain more than 700 kilometres long, north and south of Kuruman in the Northern Cape province, and in Limpopo Province and the North West province. Apart from a small output from Australia, and poor-quality deposits in Bolivia, South Africa was the sole producer of economic quantities of this group of superior asbestos minerals.

The other important asbestos mineral, chrysotile, falls within the Serpentine mineral group. Serpentine is a hydrated alteration product of igneous rocks rich in magnesia (MgO) and poor in silica (SiO_2), and fibrous deposits of this variety of asbestos are developed in rocks, notably in the Barberton area, which are amongst the most ancient in southern Africa.

Other producers of chrysotile are located mainly in the Russian Federation, Canada and Zimbabwe. In broad terms, crocidolite and amosite have higher tensile strength, resistance to acids, and binding characteristics with cement, rubber and plastics than chrysotile, which has better resistance to heat. There are other asbestiform minerals in addition to those discussed, but they are of little economic significance. A useful if brief summary of the asbestos deposits of South Africa is that given by Ehlers and Vorster (1998: 68–75), while a fuller account of the history, and medical and legal aspects has been published by McCulloch (2002).

The comparative rarity of crocidolite on a global scale is a reflection of the rather special geological conditions necessary for its formation. The so-called 'banded ironstones' hosting the seams appear originally to have been largely chemically deposited precipitates or slimes that collected on the floor of a shallow sea. The mineral, crocidolite, is a species that contains more than 5 per cent Na_2O (the oxide of sodium), appreciably higher than the 2.9 per cent which is the average for rocks of the Earth's crust. This led Du Toit (1945) to conclude that the original sediments included layers of sodium-rich clays of marine origin, within which the mineral grew during several subsequent cycles of high regional stress imposed on the rocks (that is, during metamorphism). As a result, the 'seams' in which the fibrous minerals occur are now found to lie largely parallel to the original stratification of the sedimentary layers. Within the seams themselves, the individual hair-like fibres lie, brush-like, close to perpendicular to the confining layers. Fibre lengths that were economically exploited were mostly within the range of 5 to 100 millimetres.

The first direct warnings pertaining specifically to the minute fibres would seem to date from 1898, when attention was drawn to 'the evil effects of asbestos dust' in a report issued in the UK by a factory inspector (McCulloch, 2002: 72). Evidence that accumulated through the 1920s and 1930s in the USA, Britain and Europe warned ever more insistently of the hazard associated not only with mining practice, but with that in the workplace of the factories processing the material, and even in the homes of factory workers to whose clothes the fibres clung, and in all surrounding areas contaminated by the dust. Nevertheless, there were still voices raised in defence of the industry. For example, in 1938 the Medical Officer for Metropolitan Life Assurance in the USA wrote that 'it is probable that cases of asbestosis will become uncommon' if energetic steps were taken in the workplace to control dust (McCulloch, 2002: 78).

The diseases associated with asbestos are singularly unpleasant. Asbestosis causes the build up of inelastic scar tissue in the lungs, resulting in a progressive loss of respiratory function. Lung cancer affects the lung tissue itself, while mesothelioma is an incurable and fatal cancer of the membranes enclosing the lungs. The current medical verdict appears to be that all types of asbestos fibre are responsible.

Notwithstanding the efforts to clean up the workplaces, it was an increasing awareness of the hazard of inhaling the airborne fibres that spurred the campaign against asbestos. Certainly, in South Africa, the reality of the hazard was brought home by the prevalence of lung disease amongst mine and treatment-plant workers, and reports of the affliction appearing even amongst farmers and their workers living tens of kilometres downwind from the treatment plants in the Northern Cape. A survey of villages in the Northern Cape's crocidolite-producing areas indicated in 1997 that 34 per cent of their inhabitants, with only environmental exposure, suffered from some form of pleural disease (McCulloch, 2002). Restrictions on the importation of amphibole fibre were instituted in the UK and Europe in the 1970s, and in Japan in 1988. As a result, by 1978 amphibole asbestos production in South Africa was already in decline, and in 1983 the importation of crocidolite and amosite into the UK was banned. The US Environmental Protection Agency introduced its ban in 1986.

COUNTER-CLAIMS AND WHITEWASH

Strenuous objections to the recommendations by the EPA were raised by the Asbestos Institute (AI) headquartered in Montreal, which insisted that the hazards of asbestos products had been grossly overestimated, and that there was a real possibility that substitutes (glass, ceramic and graphite fibres) would actually lead to an increase in cancer-related deaths. It was also claimed that research had established that the chief culprits were crocidolite and amosite, but that there was 'low durability of chrysotile fibres in lung tissue' (Asbestos Institute, 1986: 12). In other words, the amphibole asbestos minerals were singled out as 'the bad guys', while body tolerance towards chrysotile fibre was claimed to be much greater.

How much of all the research should be regarded as 'good science' is debatable. Competition within a multimillion dollar industry was bound to lead to some loss of objectivity. While the link between asbestos and the respiratory diseases cannot be doubted, some of the evidence that was produced to dramatise the scourge is bound to raise the eyebrows of critics. A case quoted by McCulloch (2002: 82) of mesothelioma afflicting two sisters, after having spent one morning cleaning moss from an asbestos roof 'many years earlier', would seem to demand inordinate faith in the forensic skills of medical practitioners. It should also be recognised that the identification of amphibole fibres as the prime culprits, and the implied exoneration of chrysotile fibre, was a finding that would have been more than acceptable to those other countries producing only chrysotile fibre, at a time when sanctions against Pretoria were in vogue. South Africa would, effectively, be the only country adversely affected by a ban on the amphibole asbestos minerals, and the move would weld political, corporate and economic aspirations together into a single assault.

Further aspects of research quoted by the Asbestos Institute (1986: 12) claimed that fibre lengths below 5 microns (0.005 mm) at levels of 1 fibre per cubic centimetre of air could be regarded as nearly harmless. The Asbestos Institute contended that improved industrial practice, through the use of vacuum-cleaner systems and respiratory protection in the workplace, 'could assure average exposures below 0.01 fibres per cubic centimetre'. It recommended that an exposure limit of 1 fibre per cubic centimetre, adopted by the European Economic Commission, be adopted for all forms of asbestos other than crocidolite. With regard to the latter, it recommended that most of its products be prohibited with the exception of asbestos cement, resistant seals and gaskets, and torque converters, while a limit of 0.5 fibre per cubic centimetre should in any event be applied in the workplace.

Despite the overturning of the EPA ruling by the US Appeals Court in 1991, on the grounds that 'there is no guarantee that the substitute products are less harmful and some may even constitute a greater health risk than asbestos' (Ehlers & Vorster, 1998: 68), the world now came to look askance at asbestos, and production declined drastically with decline in demand. With the disappearance of the markets, the last crocidolite mine in South Africa closed in 1997, while the production of chrysotile has fallen very significantly. Where there had been profligate use of asbestos in the construction of residences and public buildings in earlier days, the resolve to remove all traces of the material from them was estimated to have incurred expenditure of some US$40 million, up to the year 2000, on public buildings in the USA alone. An estimate for the future gross costs of cleaning up, medical care of the afflicted, and litigation, has been put as high as $500 billion in the USA by McCulloch (2002: 10). It is claimed, furthermore, that asbestos-related litigation was a major cause of the crisis of near-bankruptcy that faced Lloyds of London, the insurers, in 1993.

Legal proceedings instituted by the afflicted in South Africa are currently in progress, aimed at cash compensation for disability incurred. A report in the *South African Geological Society Quarterly Bulletin* (2003, Vol. 46: 19) noted that the mining house, Gencor, had agreed to pay a total amount of R460.5 million to 13 500 beneficiaries who claimed to have contracted asbestosis while working on or living near asbestos mines in which Gencor had stakes. The legal issues of ownership of the mines, responsibility and culpability are, however, emerging as a highly complex tangle that will take many years to resolve.

EXPLOITATION OF THE SITUATION

What has received less prominence is the level of opportunism (a word rather less emotive than the more appropriate term 'scams') that followed hysterical overreaction to the alert about asbestos. Christopher Booker, writing for *The Weekly Telegraph* of February 2003, recounts how asbestos removals firms licensed by the Health and Safety Executive in Britain have misrepresented the law to justify absurdly inflated sums for work that has been shown to be either totally unnecessary, or which could have been safely carried out at only a fraction of the cost.

A parish in Essex was quoted thousands of pounds for digging up ground in which small fragments of asbestos cement had been discovered while planting a tree. A golf club in Wales was charged £40 000 for the removal of asbestos in the roof, and then quoted another £40 000 for the removal of a further infilling of asbestos claimed to have been discovered there after the work commenced. Independent analysis subsequently showed that there had never been any asbestos in the building. In another case quoted, residents of London flats were quoted £40 000 for asbestos removal, and then another £60 000 for the removal of more asbestos claimed to have been revealed after beginning the clean-up. An independent consultant who was called in showed that the reputedly dangerous material was, in fact, horsehair.

Booker asserted that questions were to be asked in the British parliament to expose 'just how unworkable and scientifically illiterate are new regulations on asbestos management which even the Health and Safety Executive admits will cost businesses billions of pounds'. His article concludes with: 'A university-backed research programme is to expose the serious flaws in the HSE's [Health and Safety Executives'] science and show how official statistics on asbestos-related deaths are fictitious.'

26
The Namaqualand bubble of 1853-1855

'I am entirely convinced of the favourable character of these mines . . . richer and richer, and better minerals will be found . . .'
Report of the mineralogist accompanying the Simon van der Stel expedition in 1685 (Rogers, 1917: 151)

'. . . I beg to remark that for many miles around, it consists chiefly of irregular hills of granite . . . without a lode, vein or any strata: either indicative of gold, silver, mercury, zinc, lead, tin, copper, or any other metalliferous mineral.'
Capt. John Davies (1846), a mining agent from Cornwall (Smalberger, 1975: 20)

'It appears that . . . [the mining agent from Cornwall] may have seen *a* mine, but certainly not *the* mine.'
Honorary Secretary (1846), South African Mining Company (Smalberger, 1975: 20)

'. . . we may come to the conclusion that this country is destined to produce annually more copper than all the mines in the world united.'
French Consul at the Cape in 1855 (Smalberger, 1975: 35)

The 1685 expedition of Simon van der Stel, Governor at the Cape, to Namaqualand (the arid, mountainous north-western part of the Cape province) to search for copper deposits rumoured to exist there is a well-known early foray into exploration for mineral deposits in this country. The search was aimed at upgrading the significance of the 'Cape of Storms' from that of a mere port of call for the ships of the Dutch East India Company taking on board vegetable and fruit supplies during their long voyages to the East. The story of prospecting in the following 300 years, flotation of companies to exploit the copper ores, and eventual successful mining here, is told by Smalberger (1975) in his *History of Copper Mining in Namaqualand.*

The report by Frederich von Werlinckhof, a mineralogist who accompanied Van der Stel's expedition (quoted previously, from Rogers, 1917), was eventually to prove prophetic, but the harsh climate, the hostile environment, shortage of water and fuel, and formidable transport problems put a damper on any serious activity for the following 150 years. It was not until 1836 that an expedition mounted by a James Alexander started a train of events that led to the floating of the first enterprise, by the South African Mining Company in 1846. The published rebuff by Captain John Davies of the claims made in the prospectus (quoted previously) would seem, however, to have scared off prospective investors, and, despite the finding of exceptionally good ore, the market value of the shares dropped to zero within two years (Smalberger, 1975).

It was the acquisition, in 1850, of the mineral rights to approximately ten hectares of the farm Springbokfontein for £50, and then subsequent purchase of the entire farm for a further £2 345 by the firm of Messrs Phillips and King, that started sustained commercial operations in copper mining in Namaqualand. Starting with an export of 11 tons of ore in 1852, the output of the Springbok Mine was to reach 2 100 tons by 1857. This, together with a proclamation relating to the issuing of leases for mining of metals other than gold in Namaqualand in 1853, was the spur that started an unprecedented but largely fraudulent boom in the floating of one company after another, and wild speculation in shares between late 1853 and early 1855.

Within a period of seven months preceding January 1854, some 22 companies were formed, and by late 1854 this number had risen to 30 to 40 (Smalberger, 1975: 46). The market value of shares soared to eight to ten times their paid-up value. Parcels of shares were reported to have changed hands up to ten times per day, with profit being made on each transaction. From Smalberger's account, it is clear that the investments made by the gullible public were from the beginning largely destined for the pockets of the promoters of most of the schemes. Many of the companies held no mining ground or leases at all and had no intention of acquiring them, and their 'agents' had no knowledge whatsoever of geology, the deposits of the region, or the logistical problems raised by limited resources of water and fuel, or transport across the terrain.

While the region erupts in an extravagant display of colourful wild flowers after sparse spring rains, it is for most of the year an arid place scorched by the Sun in summer, and swept by bitterly cold winds during winter. Living conditions are harsh. Regrettably, a report by the respected Andrew Geddes Bain was much quoted by the carpetbaggers to stoke the euphoric fires, for he had written that 'Namaqualand contains more mineral wealth than the whole of the Colony put together'. On top of all of this, a baseless rumour to the effect that payable gold had also been found kept the pot boiling merrily.

What happened next was inevitable – the bubble burst. Shares had been bought using bills that were never honoured. With the exception of three, the companies were all liquidated and the speculators disappeared with the cash. There remained only Messrs Phillips and King, the Namaqua Mining Company, and the South African Mining Company (which survived only briefly).

The irony of this entire episode was that there was, indeed, solid mineral wealth waiting to be exploited (see Plate 18). In the initial stages, the costs of transporting the ore from Springbok Mine had militated against shipping any consignment averaging less than 25 per cent copper metal – a fabulously high grade by the standards of today. By 1873, O'okiep Mine had reached a depth of 130 metres and was yielding ores with 30 to 60 per cent copper, and lower-grade material that was upgraded by various ore-dressing processes. Mining in the region continued until 1931, by which time it had produced 2.2 million tons of ore averaging 14 per cent copper (Lombaard et al., 1986).

In 1937 the O'okiep Copper Company acquired all the assets in the region, and by 1979 some 67 million tons of ore had been milled, and the estimate was made that total production since 1852, and remaining ore reserves, accounted for 2 million tons of copper metal. Persistence to considerable depth of some of the ore bodies had been proved – in the case of Carolusberg Deep, to a depth of 1 500 metres (Marais et al., 2001). The estimate for total copper recovered plus remaining ore reserves in the region, on a basis of 2005 prices for copper metal, would place the total value of this at some US$7 billion. Impressive as this figure might be, it falls far short of the wildly optimistic prediction in 1855, by the French Consul at the Cape, that Namaqualand would yield more copper than the rest of the world combined.

The reasons for this unwarranted optimism in the earliest days are not difficult to find. They are rooted in the geology of the region. First impressions of this suggest to the casual visitor that here is a sequence of gently folded gneissic rocks into which has been intruded a suite of igneous rocks (formally known as the Koperberg Suite) which bear the copper minerals. It was the mining of these bodies that has yielded the output. Between the years 1946 and 1984, mapping, drilling and geophysical exploration by the O'okiep Copper Company had located no less than some 1 700 of these intrusive bodies, of which 320 were classified as 'copper prospects' (Marais et al., 2001), with few being totally devoid of copper minerals.

The world over, prospectors and miners have learned to their cost the bitter lesson that 'a little copper goes a long way' in producing showy surface deposits of the brightly coloured carbonates (see Plate 15), malachite (green) and azurite (blue). An entirely false impression of the size of the deposits at depth may be given. It is not surprising, therefore, that early visitors such as Bain conjured up a picture of unlimited reserves of the metal underlying a huge area, rather than a profusion of separate pods, the majority of which do not warrant mining.

It seldom happens that the first optimistic impressions relating to a particular mineral deposit are fully substantiated in subsequent years. Von Werlinckhof would therefore be deeply gratified, were he able to rise again like some latter-day Rip van Winkel, to see the words that he had penned in 1685 in juxtaposition with those in the most recent publication by Marais et al. Von Werlinckhof had declared that it was '. . . certain and obvious that the deeper one goes the richer and richer the minerals will be' (Rogers, 1917), while Marais et al. stated that the '. . . assay plans of all the mines developed . . . unquestionably demonstrate the downward enrichment of copper tenor' (2001: 39).

27
The chequered history of the Cape ruby

'The beautiful red stones that are associated with the diamond in South Africa were almost a drug on the market [an unsaleable commodity] under their proper title, garnet, but commanded a ready sale under the misnomer "Cape Ruby". . . it really belonged to the cinderella species of gemstones, the despised garnet.'

G.F.H. Smith (1940: 40)

To experience the ultimate tangle in the giving of names, we need look no further than the world of gemstones, where there is a bewildering array of misleading terms that originated, for the most part, in the jewellery trade. 'Brazilian sapphire' is in reality blue tourmaline, 'Indian topaz' is yellow corundum, and green fluorite was once known as 'African emerald'. The 'Cornish diamond' is colourless quartz (see Plate 17), the 'Ceylon diamond' is zircon and the 'Saxony diamond' is topaz. Also in this bag is the 'Cape ruby', which is not a ruby, but a member of the Garnet Group.

Quite apart from rubies and garnets being quite distinct species from a mineralogical standpoint, the difference is emphasised by the price that dealers have been prepared to pay for them. Herbert Smith (1940) put this in perspective in his book, *Gemstones*, in which he quotes the price of four-carat rubies in the earlier part of the twentieth century as £200 per carat, against a humble 5 shillings per carat for the Cape ruby. Our focus here is not, however, to decry the misnomers in the jewellery trade, but to illustrate how a somewhat insignificant red crystal that often rather closely resembles a red currant has influenced ideas about the structure of our planet.

Six main species make up the Garnet Group – all hard minerals of high density, whose usage ranges from their being cut as semi-precious gemstones to the more prosaic manufacture of sandpaper used by cabinetmakers. The six species originally took their name from their supposed resemblance to the luscious fruit of the pomegranate tree. In times long past, they were more highly favoured than they are today, and ranked, for example, alongside diamond, ruby, sapphire and emerald as one of the 'Nine Gems' that were the symbol of kingly status in Thailand (Wannenburgh & Johnson, 1990).

Of the six common species in the group, it is *pyrope* that we know as the Cape ruby – a magnesium–aluminium silicate – $Mg_3 Al_2 (SiO_4)_3$ – that is recovered in abundance as chips and crystals during the search for diamonds. It is one of the 'indicator minerals' that prospectors look for, together with the black titanium mineral, ilmenite, and the green crystals known as chrome diopside. The reason why it is so abundant here is that it is carried within the rock that is the host of the diamond – the rock-type known as kimberlite.

A GEOLOGIST'S BAROMETER

What is generally less well known about pyrope is the role that it played 100 years ago in the development of concepts of the structure and history of our Earth. This came about when it was recognised that pyrope can form only in environments of high pressure, or, in terms of Earth structure, at deep levels of the crust and in the rocks of the mantle that underlie it. In 1913 Sir Lewis Fermor (see Fermor, 1938: 49) wrote that 'garnet [pyrope] is unknown as a product of solidification of fluid magmas under atmospheric pressure', and that it 'could be regarded as a geological barometer'.

Pyrope is produced in nature where several very common rock-forming minerals (feldspars, pyroxenes and olivines) are forced by high pressures to react to form a new, more dense phase – garnet – with a shrinkage of volume by some 10 to 20 per cent. In this way basalt, a rather nondescript black or grey lava, undergoes under pressure a metamorphosis into a magnificently handsome rock of red pyrope garnet intergrown with a green pyroxene, omphacite (a name stemming from the Greek word for 'unripe grape'). This product is known as eclogite, or, as it was once named in the case of fragments of it brought up from the depths in the diamond pipes in South Africa, as griquaite.

Amongst the most widespread of all rocks building the Earth's crust is the aforesaid basalt, forming the floors of the oceans, and covering huge areas as lava fields in Russia, western USA, India, Africa (for example, the Drakensberg Mountains) and elsewhere. It also has a coarser-grained equivalent, gabbro, which probably makes up a significant proportion of the deeper levels of the crust beneath the continents. It was this superabundance that led Fermor in 1913 to remark that 'the conclusion accepted by almost everyone . . . is that there must be a continuous shell of rock which, when melted, forms the basaltic lava that ascends through cracks to the earth's surface. We are hence driven to the acceptance of the idea of a continuous basaltic shell to the earth' (1938: 60).

From his work on garnets in the rocks of the subcontinent of India, Fermor took the seemingly logical next step of reasoning that at depth, where pressures are high enough, rocks of the basalt clan would have to be represented by their high-density equivalent, eclogite. This metamorphosis would start at a depth of somewhere between 35 and 60 kilometres below surface, and this, he asserted, would everywhere involve a sudden increase in the density, or specific gravity from around 2.7 to 3.3. Starting with this premise, he then went on to offer ingenious solutions for a number of pressing geological problems.

FERMOR'S FIRST PROPOSITION: HOW LAVAS FLOW GREAT DISTANCES

Basaltic lavas are extruded from volcanoes or fissures at bright red heat, but cool rapidly when they come in contact with air, rain and the cooler surfaces over which they flow. Puzzled by the extraordinary distances that some lava flows have spread (an example of a flow advancing some 100 kilometres was quoted) Fermor had concluded that this could be possible only if the lava had in the first instance been extruded at exceptionally high temperatures. His conclusion was that some basalts were erupted with as much as several hundred degrees of superheat – that is, at temperatures greatly in excess of that required merely to achieve melting. This superheating he attributed to the melting of eclogite at depths exceeding 60 kilometres, to yield magmas at temperatures appreciably higher than those needed to melt basalt at shallower depths.

FERMOR'S SECOND PROPOSITION: THE MATERIAL UNDERLYING THE CRUST

We have for millennia wrestled with the question of what lies beneath the crust on which we live. It is certainly not water, as was believed by Revd Thomas Burnet in 1681, nor was it the 'cold Earth' of Ernest Schwarz (see Chapter 4). Since we cannot hope ever to get down there to see for ourselves, an answer can be sought only in indirect evidence. Fermor's answer to this was that there must be a world-encircling layer made of the heavy, colourfully red-and-green eclogite at depth.

Geophysics seemed at that time to back up what Fermor had surmised. The seismologist, A. Mohorovicic, had in 1909 discovered the worldwide discontinuity (now known in his honour as the Moho) at the base of the crust, where there is a sudden increase in the velocity of seismic (earthquake-generated) waves propagated through the rocks. The data fitted the properties of eclogite. This concept of a worldwide layer of eclogite was an appealing one that lingered in the minds of geologists for many decades, and, even as late as 1965, one of the world's leading geologists, Arthur Holmes, was still undecided about its role in defining this discontinuity.

Holmes had conceded that the melting of eclogite could provide a significant source of basaltic magma (1965: 274, 1246). Furthermore, the eclogite fragments found in kimberlite pipes prompted his comment that: 'we are clearly given a broad hint that layers or bands of eclogite . . . are likely to be everywhere present at some depth . . .'

In so far as Fermor's concept was based on what appeared to be sound argument, it was good science. Unfortunately, he turned out to be wrong. The controversy that flared up was whether this discontinuity, the Moho, was produced by a change of physical state (as in the transformation of basaltic rock to high-density eclogite, without change of composition), or by a change to a rock type of totally different composition, which many believed to be peridotite. The peridotites make up a family of dark, heavy rocks with abundant olivine and pyroxenes, with or without garnet, and the physical properties of eclogite and peridotite are so similar that geophysical methods failed to distinguish between the two.

It fell to A.E. Ringwood, Professor of geochemistry at the Australian National University, to disentangle the controversy. In his *Composition and Petrology of the Earth's Mantle* (1975), he summarised the results of several hundred experimental runs in the laboratory that showed that the transition from basalt to eclogite was not as abrupt as would be demanded to account for the well-defined Moho Discontinuity. The transition was, rather, smeared across a wide band equivalent to a vertical depth range from 10 to 35 kilometres wide. This did not fit the model of an abrupt change at the Moho in any way. The material underlying the Moho was deduced to be peridotite, and further experimentation over a period of decades has yielded compelling evidence that it is partial melting of this that yields the voluminous basaltic magmas. Accordingly, the discontinuity at the Moho is not to be regarded as a change of state to eclogite, as Fermor believed, but as one to a rock of very different chemical composition.

The deep diamond mines indicated the same thing: statistically, eclogite averages a mere 5 per cent of the exotic fragments in them, brought up from the mantle by the rising magma, while peridotite fragments, or rocks closely related to them, make up the remaining 95 per cent. There remained little persuasive evidence for retaining Fermor's concept of a worldwide layer of eclogite.

FERMOR'S THIRD PROPOSITION: THE RISING AND SINKING OF CONTINENTS

Studies of ancient shorelines have shown that their survival at different heights is due to two major factors – the changing volume of water in the oceans, and upward and downward movements of the continents themselves, attributable to isostasy. This is the mechanism that causes land masses to rise where erosion strips off the upper parts of the crust exposed to attack, or causes them to sink if they become loaded with thick accumulations of sediments or ice.

Fermor had, for example, observed that despite the rapid accumulation of sediment in deltas such as that at the mouth of the Ganges River, the surface of the delta sediments remained at a constant level but a few feet from sea level. To account for this, he invoked the transformation of underlying basalt to eclogite at depth, with a reduction in volume, as the crust became loaded at surface. If, at other places, erosion were to be the dominant process, or if a thick ice cap should melt away, the loss of mass at the Earth's surface would lead to an expansion of volume where eclogite reverted back to basaltic rock. In other words, he saw the basalt–eclogite transformation as a form of shock absorber that always took up the slack, or responded to increased loading.

GARNETS AS THE BIRTH CERTIFICATES OF THE DIAMOND

From the way the majority of kimberlite pipes in South Africa are drilled through relatively young strata, it is known that they were intruded only ~100 million years ago, or less. The logical inference would seem to be that this would be the age of the crystalline materials that make up the pipe-filling that cooled and crystallised there. Research within the last few decades has, however, brought to light the surprising revelation that this is not so – the garnets associated with the diamonds in kimberlites may be of extreme antiquity.

S.H. Richardson and his colleagues at the University of Cape Town (1984, 1986, 1990) have used isotopic systems to establish that here the garnets have been formed at various times between 3 300 and 990 million years ago, and that two types of garnet exist. One variety stems from the deep-seated peridotites that yielded what are known as P-type diamonds (P for peridotite), and the other from eclogites hosting E-type diamonds (E for eclogite). This means that the little red garnets had been lying dormant in the mantle for some 900 to 3 200 million years before being rudely picked up by ascending kimberlite magma and carried towards the surface. This in turn means that diamonds must occur in profusion in the Earth's upper mantle, but we will never have direct access to them there – they lie at depths of more than 150 kilometres below surface, a long way below our deepest mines.

What Herbert Smith had referred to as the 'despised garnets', in the quotation at the start of this chapter, may not be of much intrinsic value, but their contribution as a spur to reaching a better understanding of the structure and history of our Earth has been profound. The sad truth remains, however, that none of this was reflected in the selling price of a little more than 5 shillings per carat.

28
DIGGERS' TESTS FOR DIAMONDS

An old diamond digger's test for a diamond was to weigh it carefully, clamp it in a vice or pliers, and file it vigorously with a whisky bottle for several minutes before weighing it again. If it was found not to have lost weight after this, it was pronounced a diamond!

'Test' used by early prospectors in this country.

The above test is worthless. All this would show is whether the 'gem' being tested is harder or softer than bottle glass. A diamond's ability to scratch glass is shared by many other minerals that will do the same, including one of the commonest on the surface of the Earth – quartz. The test is therefore about as useful as the medieval belief that brown diamonds were male and colourless stones female, and that small diamonds were produced by the mating of the two.

It is ironic that journalists repeatedly refer to the test of 'scratching a window-pane' during the early days when the first diamonds found in South Africa were passed around. The initials 'JR' of Father James Ricards, scratched in 1867 on a glass pane in Grahamstown, where the first authentication of a stone picked up in the Hopetown district was made by Dr W.G. Atherstone, are a matter of historical record. The glass pane still exists (and a photograph of it appears in Metrowich [1953]). The implication that this was the definitive test used by this medical doctor and amateur geologist is a slight on the competence of a remarkable man.

What is not generally known is that Atherstone was entitled to write the letters MD, FGS, FRGS after his name, and was elected Honorary Vice-President of the Geological Society of South Africa in the year of its foundation in 1895. The biographical sketch by Schwarz (1908) recounts that, amongst other tests, Atherstone had employed a polariscope, tested the stone's hardness against sapphire, and determined its specific gravity as 3.5343. Confirmation of his identification, according to Schwarz, came later from Messrs Hurst and Boskell, jewellers to the Crown in London. This stone, subsequently named the *Eureka*, weighed in at 21.19 carats.

It would seem unlikely that a narrative as simple as this could give rise to a passionate denunciation of the man who penned it. Marian Robertson's *Diamond Fever* (1974) contains a meticulous documentation of the events of 1866–69, transcribed from primary sources. In it she recounts that Mrs Minnie Cartwright, Atherstone's granddaughter, had taken umbrage at what Schwarz had published, especially so in view of the wide credence that his account was subsequently given. She averred that Messrs Hurst and Boskell had not been the jewellers who confirmed the identification of the first stone, but Messrs Garrard and Co., while Hunt and Roskell (not Hurst and Boskell) was the company that had handled 'The Star of South Africa'. On Atherstone's receipt of the letter containing the diamond, she added, the stone had not dropped from the envelope to the floor of the Post Office, where, according to Schwarz's account, it was nearly lost, but had fallen into the grass where Atherstone had been sitting in a chair outside his home when he opened the letter. Trivial details indeed, but worth recording in view of the historical significance of the event.

Of all the tests ever applied to identify a diamond, the most bizarre must surely have been that used by Messrs Nicholson and Cruikshanks, a pair of traders working along the course of the Vaal River at the time when the first diamonds were being recovered in South Africa. Robertson (1974: 126) quotes from papers lodged in the McGregor Museum in Kimberley that these two had prepared a fire and then roasted in it a diamond of four carats recovered in the very early days of the diamond rush. Only after this misguided 'test' had they pronounced themselves satisfied that it was indeed a genuine diamond. But the damage had already been done. The account goes on to say that the test had produced black spots in the stone, and that these blemishes had degraded its valuation to a mere £20. Nicholson and Cruikshanks would have done better to touch the stone to their tongues or foreheads. The extraordinary capacity of the diamond to conduct heat rapidly away from any surface with which it is in contact produces a sensation that the stone is always cold.

'DIAMONDS ARE FOREVER'?

The combination of properties found in diamonds is unique – hardness, refractive index giving them their brilliance and sparkle, ability to be polished, thermal conductivity, use in industry, etc. – all of which makes them objects of intense desire, but diamonds are, in fact, far from being indestructible. At high temperatures in the presence of oxygen they burn off to carbon dioxide. A delightful but startling quote from Dana's *System of Mineralogy* (1904: 4) records that in the absence of oxygen a diamond 'is transformed into a kind of coke'. Although the hardest natural substance known, it is brittle, and easily cleaved by an expert into smaller pieces along specific crystallographic planes. The mistaken belief that the diamond is indestructible has led to many a gem being destroyed by 'testing' it in a vice or pliers, or pounding it to dust with a hammer wielded with vandalistic enthusiasm.

An investigation into the stability of diamonds was reported in

1991 in the *Mintek Bulletin*, under the heading 'Diamonds Are (Not) Forever', from which the following extract is taken: 'All diamonds are unstable in air, and it is only because of the extremely slow kinetics of transformation that the stone has become a symbol of permanence . . . diamonds react rapidly at high temperatures in an oxidising environment.' Under experimental conditions, etching and flaking of surfaces is evident at 1 000 °C.

Another peculiarity of the diamond is a proclivity towards self-destruction. Numerous recorded cases tell of stones falling apart or even 'exploding' on being taken from the 'blue ground' in which they were mined, or during subsequent cutting. This might happen immediately after recovery, or cracks might develop only years later. Some buyers used to embed the stones that they bought from the diggers in a potato so that, if the stone should disintegrate, the fragments would not be lost.

Old beliefs that exposure to dry rather than moist air, or to the sun, would encourage spontaneous cracking, were long ago dismissed by Alpheus Williams (1932: 426). He did note, however, that cracking of stones is usually centred around foreign inclusions held within the stone, particularly in the case of the inclusion of one diamond within another, when even a gentle tap was said to be adequate to break the outer crystal. The use of polarised light shows that diamonds are under stress around inclusions of a variety of other trapped minerals, or cavities within the crystal. This need occasion no surprise, as differences in coefficients of expansion between host crystal and inclusion would inevitably set up internal stresses on ejection from the high-pressure, high-temperature environment in which initial growth of the diamond took place.

29
Diamonds in South Africa: Some bad early calls

'. . . are the diamonds said to be found at the Cape, and sent to England, really found in South Africa? . . . The geological character of that part of the country renders it impossible . . . that any could have been really discovered there . . . I can now only conclude by expressing my conviction that the whole diamond discovery in South Africa is an imposture – a Bubble scheme.'

James R. Gregory (1868a), mineralogist

'The whole thing [the prospectus for offering shares in the newly discovered Premier Mine] was a fake.'

Judgement of the Chairman of the De Beers Board of Directors in 1902

(Kanfer, 1993: 159)

'. . . the chances of locating a rich pipe outside of . . .[the tract of country between Jagersfontein in the Free State, and Premier Mine east of Pretoria] . . . are evidently very small . . . [however] Africa has ever been a land of surprises.'

Percy A. Wagner (1914: 4), consulting geologist

Thes preceding few extracts from early sources illustrate the dangers of making sweeping predictions in Earth science. As it turned out, each of the predictions could scarcely have been more wrong.

James Gregory was a mineralogist who was sent out to South Africa by a Mr Harry Emanuel, a diamond merchant of Bond Street, London, shortly after the first reports of the finding of diamonds in the Cape had hit the streets. But for the insinuations of duplicity that subsequently surfaced, history might not, perhaps, have treated Gregory so harshly for his aspersions as up to that date diamonds had been found only in alluvial deposits, most notably in India and Brazil, and his approach had been to compare the local terrain, and the gravels he saw there, with those of the other known sources. He claimed that he had found no similarities on which to base an optimistic forecast, but then stepped from reality into fantasy in suggesting that diamonds had been carried to Hopetown by wandering ostriches.*

It was Gregory's misfortune that it was only in the following year that diamonds were then found in inconceivable quantities within a hitherto unknown rock type – *kimberlite* – first at Jagersfontein and then Kimberley. Within five years of his visit, South Africa was producing more than 90 per cent of the world's diamonds. The literature does not record what words about this passed in private between Emanuel and Gregory, but on 6 January 1869 the *Eastern Province Herald* reprinted from the *Journal of the Society of Arts* in London a letter from Emanuel written in the previous November (Robertson, 1974: 162). Here Emanuel flatly rejected the possibility of diamonds being found in South Africa, using emphatic words to scare off any would-be adventurers who believed the tales, because 'some unfortunate persons may thereby be induced to embark on a fruitless errand'.

Emanuel insisted that Gregory had carefully visited the Orange, Vaal, Buffalo and Fish Rivers during a journey of over 3 200 kilometres, to arrive at his conclusion that 'no diamonds have, or ever will be found in the Cape Colony – saving such as are deposited there for a purpose . . .' Emanuel's letter is peppered with gross inaccuracies, such as the claim

* Ostriches are well known for their habit of swallowing pebbles, and, as any bejewelled visitor to an ostrich farm will know, they will peck at any bright objects such as earrings, brooches or necklaces. In his book *To the River's End*, Lawrence Green (undated, c.1950) tells how Hendrik Louw, a hunter and farmer close to the Orange (Gariep) River mouth in the late nineteenth century, shot 'thousands of ostriches in the old days' when the feathers fetched £36 per pound. He often found 'bright stones' in the birds' stomachs without recognising them as diamonds, thereby missing a fortune under his nose. Nevertheless, it would seem that Gregory did not ask the logical question: where would his ostriches have found the stones in the first place? Ostriches are flightless birds, and could hardly have flown here from Brazil or India.

that the whole of the territory explored is of volcanic origin, except for the granite at the hamlet of George, and that, hitherto, diamonds had been found only among 'rocks of the very earliest period, such as granite mica slage [*sic*] and other metamorphic rocks'. There are, in fact, no volcanic rocks within hundreds of kilometres of the George, Buffalo and Fish River areas, apart from a tiny outcrop in the Zuurberg Mountains near Port Elizabeth. Why Gregory had been scratching around these areas remains a mystery: they are upwards of 600 kilometres distant from where the diamond fields were, and Gregory had been instructed to 'report', not to prospect. It is not therefore surprising that insinuations were then made that Gregory had been sent out to South Africa specifically to deliver a damning report, in order to keep up the market price of diamonds in London (Kanfer, 1993: 25).

In the *Grahamstown Journal* of 23 January 1869, G.F. Gilfillan asserted, after tracing Gregory's movements in South Africa, that he had never been near the diamond fields. Finally, in May 1869, Emanuel was obliged to respond by letter to the Colonial Secretary, Richard Southey, in which he conceded that 'I can no longer have any doubt of the fact of Diamonds being found in your Colony', but disclaimed any responsibility for misleading the public, insisting that 'I cannot consent to bear the onus of any misstatement on his [Gregory's] part.' Robertson (1974: 204) quotes further from a letter written by Atherstone in August, in which he slates Gregory for 'total want of principle and common honesty'. Perhaps the full truth of the strange episode will never be known.

The Chairman of the De Beers Board of Directors at that time, as quoted at the start of this chapter, was not alone in his hasty judgement of the future potential of the Premier Mine. Alpheus Williams (1932: 25) recalls that the Government Mining Engineer of the Transvaal had been equally dismissive in 1903. In his report the latter had written, 'I am not aware, however, that any instance has been recorded when the blue ground [of the diamond mines in the Pretoria district] has been proved payable.' Both the Chairman of the Board and the Government Mining Engineer were grievously mistaken. By 1932, Premier Mine had recovered well over 28 million carats, and was famous for its yield of stones that were of exceptional size and quality. A quarter of the world's diamonds exceeding 400 carats have come from here, including the largest ever found – the Cullinan Diamond weighing in at a massive

3 106 carats, from which the Great Star of South Africa (530 carats) in the British Crown jewels was cut. As massive as the Cullinan diamond was, it was firmly held by many that it was merely a fragment broken off an enormous stone more than double that size. If the other half does exist, it has never been found.

Given the determination of both prospectors and mining companies to smell out the precious gem, wherever it might lie hidden, the prediction of Percy A. Wagner in 1914 (quoted at the start of this chapter) was also to prove a rash one for a man of his eminence and experience, even if he did allow that Africa was a land of surprises. The discovery of diamond-bearing pipes in Tanganyika (Tanzania) was highlighted by operations at Mwadui Mine, south of Lake Victoria, where the daily target during the 1950s was 1 000 carats. The Orapa Mine of Botswana, and Finsch Mine in the Northern Cape province, have been prolific producers. Also in Botswana is Jwaneng, the richest pipe ever discovered. The Venetia Mine in the Limpopo Province, just south of the border with Zimbabwe, was opened in 1991, and was by 1995 yielding 4.3 million carats per annum, at an average grade some three to four times higher than was typical of the mines of the Kimberley area in their first twenty years of operation.

30
Where do diamonds form? The beginnings of a 100-year dispute

'I think I am justified in saying that if the diamond problem is not actually solved, it is certainly no longer insoluble.'

William Crookes, FRS (1897b)

The evolution of theories on the genesis of the diamond offers a narrative as fascinating as any in the annals of Earth science – not because of the value placed on the gem, or ancient myths about its supposed mystical properties – but in the unravelling of a complex geological history from the combined hints provided by geology, physics and chemistry.

Admittedly, there have been some who have decried the value put on the stone in science and commerce: John Ruskin (1880: 54) gave his provocative opinion in the rather absurd comment: 'The worth of a diamond is simply the understanding of the time it must take to look for it before it is found . . . for a diamond has no more real beauty than a piece of glass.' Happily, there are not many who would agree with this dismissive rejection of what Marilyn Monroe's famous song hailed as 'Girl's best friend'.

THE ALLUVIAL DIAMONDS OF INDIA AND BRAZIL

Prior to the discovery of diamonds in a dirty-green, soft and decomposed igneous rock that Professor Carvill Lewis had named *kimberlite* in 1886, the diamonds of the world had been recovered from river gravels, first in India and then in Brazil. While the wearing of diamonds as ornaments goes back, according to Wannenburgh and Johnson (1990: 20), some 4 000 years, there is no reference to diamonds being one of the twelve gems set into the breastplate of the High Priest of the Israelites in the story of Moses (Exodus 28). There was probably a good reason for this. As Herbert Smith (1940: 162) has pointed out in his authoritative treatise on gemstones, the hardness of the stone is so much greater than that of any other natural material that the stipulation that each stone of the breastplate should be inscribed with the name of one of the Twelve Tribes could not have been met at that time.

The Arab and Persian trade in Indian diamonds was already well established by the fourth century BC, and certain Indian diamond-trading ports were well known to the Greeks in the first century BC. As late as the sixteenth and seventeenth centuries AD, Golconda in the state of Hyderabad was still the 'diamond capital' of the world, a place where the stones recovered from the tributaries of the Krishna River were traded and from there fed into world commerce. Only in 1726 was this monopoly broken by new discoveries in the Minas Gerais district of Brazil.

All known sources up to 1870 were alluvial deposits that had trapped the diamonds shed by the weathering and erosion of older but unidentified and untraced host rocks that were the primary source. The myths of the Ancient World had it that diamonds there were born from an act of union between male and female stones while in the gravels. This myth probably originated in the occurrence of some natural gems, known to crystallographers as twinned crystals – an intimate interlocking of separate parts of what resembles a pair of crystals symmetrically facing each other.

AT LAST – DISCOVERY OF THE PRIMARY SOURCE

It was in 1870 that the first discovery of the stones embedded in the enigmatic but undoubtedly igneous source rock at Jagersfontein, in the Orange Free State, finally burst upon the world, starting a scientific controversy that raged for more than 100 years before some degree of consensus about the nature of kimberlite was reached late in the twentieth century. This does not necessarily mean that all the problems surrounding the source, and subsequent spreading of diamonds, and their capture within gravelly deposits, have now been resolved. Some of these problems still linger with us today.

The rock type known as kimberlite is bluish-green to grey when fresh, but yellow-brown where weathered to crumbly 'yellow ground' immediately beneath the surface. With a little experience, the dominant minerals are easily recognised – greenish olivine (peridot), greenish-grey orthopyroxene, apple-green chrome-diopside, red and purplish garnets, golden-coloured mica, black and opaque oxide minerals ilmenite, magnetite and chromite, and soft carbonate minerals. Serpentinisation, the process of breakdown to a soft, green alteration product, is generally conspicuous. Scattered within this blue or yellow 'ground', at a level of about 1 part to every 5 to 15 million parts of kimberlite host rock in workable deposits, is the diamond. This dwindles down towards zero in unpayable pipes.

From the outset, it was realised that here was a rock that had had an unusual and complex history. Sir William Crookes, FRS, had noted in his much-quoted lecture to the Royal Society (1897a) that more than 80 different species of minerals had been found within it, and commented that 'a more heterogeneous mixture can hardly be found anywhere else on this globe'. (A rather more irreverent description has been used by others who have described diamond pipes as 'Nature's garbage cans'.)

Also embedded within the greenish or greyish kimberlite matrix were abundant fragments (loosely termed 'boulders' or 'nodules') of coarsely crystalline rocks that bore the stamp of having originated at great depths below the surface before being disrupted and caught up in hot kimberlite magma thrusting its way towards the surface. During this long journey of escape, these fragments had become corroded, abraded, rounded and even polished, to resemble water-worn boulders that might be found along any river course.* The term *xenolith*, deriving from Greek (*xenos* meaning strange), was applied to these fragments, implying some exotic source. From early on, it was recognised that two families of rocks were represented here – one the handsome rock composed of green pyroxene studded with red garnet, known as *eclogite* or *griquaite*, and the other a varied assemblage of rocks rich in either the pale-green mineral, olivine (the *peridotites*) or pyroxene (the *pyroxenites*), or both.

* This was, in fact, an interpretation offered by Professor Bonney in 1899, and reiterated by him in 1907 in the words: 'they are water-worn boulders, which were once part of a conglomerate, whence they have been scattered by the volcanic explosions . . .'

A further component of the population of these xenolithic fragments in kimberlite included other rocks of very different type, recognisably torn from near-surface strata through which the invading kimberlite had streamed during its forceful rise to surface (granites, schists, lavas and sediments). Included with these was yet other detritus that could only have been derived from strata that have now disappeared from the scene – stripped off the surface and carried away by erosion through more than 50 000 millennia. Fossils in these latter xenoliths, fragments of coal, and fossil trees characteristic of strata that elsewhere occur thousands of feet higher up in the Karoo Sequence, tell of their presence in past times.

From these features arose the concept of turbulent emplacement of kimberlite which, at any one place, might contain a chaotic assemblage of disrupted fragments – those carried up from very great depths being juxtaposed with others from relatively shallow depths. The turbulent mixing that was pictured could be likened to the tumbling, churning motion of sand and gravel in a concrete mixer.

A puzzling feature of the kimberlite exposed in the mines, despite its deep-seated origin, was the absence of any indications that it was intruded at high temperature. As Crookes (1897a) recognised more than 100 years ago, 'the blue ground does not show any signs of passing through great heat, as the fragments in the breccia are not fused at the edges.' Clearly, there was a paradox here. The pipes themselves showed all the attributes of volcanic eruption, but this was belied by the seemingly low temperatures of the pipe contents.

With regard to the mode of occurrence of these kimberlites, it was known from mining that they occupied pipe-like or carrot-shaped repositories drilled through the crust, tapering downwards and ultimately shrinking into narrow, wall-like dykes through which the magma had once streamed. Similar narrow, dyke-like bodies, some richly endowed with diamonds, were known to reach the surface (see Plate 20).

Thus was the stage set for the great dispute or, more correctly, the opening of a whole can of scientific worms. Where did the diamonds come from? What was the source of the carbon of which they are made? Did diamonds form in the great depths in association with the rocks now represented by the xenoliths of deep-seated origin, or did they crystallise within the kimberlite magma during its ascent towards surface? If so, when had this taken place? These were fundamental questions that focused on the genesis of the diamond itself.

Then, there was a second category of questions that asked how and when the kimberlite was emplaced within the pipe-like conduits reaching the surface. Was the act of emplacement characterised by violent, explosive activity, or was it an essentially tranquil event? How often was the injection of kimberlite repeated within the same pipe, or was there but a single event? Why did adjacent pipes carry diamonds that may be strikingly different in size, quantity, colour (colourless, blue, brown, yellow, green or black), and shape (sharp-edged octahedra, rounded forms, pitted or grooved crystals, and broken fragments, etc.)?

It was not until the later years of the twentieth century that some clarity began to emerge from the scientific work that was to follow the first discoveries.

31
KIMBERLITE AND THE GENESIS OF THE DIAMOND

'It is conclusively proved that the diamond has not been formed in situ in the blue ground . . . genesis must have taken place at great depths under enormous pressure.'
Sir William Crookes, FRS (1897b)

'The fragmental character of many of the diamonds is as explicable on the assumption that they were formed in the molten magma and broken during the movement which went on in the pipes, as it is by assuming that the diamonds are derived from some foreign rock.'
F.H. Hatch and G.S. Corstorphine (1905: 299)

'I think we may dismiss from consideration any such ideas as their [the garnet's] derivation from churned up eclogite . . . I think [this] is quite sufficient to disprove also the theory that the pipe diamonds come from eclogites at all.'
F.P. Mennell (1908)

'. . . kimberlite magma was generated by the liquefaction . . . of portions of a universal zone . . . of . . . peridotite, which, at some unknown depth, must be assumed to underlie the platform of granite and gneissoid rocks forming the basement of Southern Africa . . .
The diamond is an original constituent of kimberlite.'
Percy A. Wagner (1914: 117, 160)

'It is generally accepted today that diamond has never yet been artificially formed and . . . its production in the factory or the laboratory is unlikely in the extreme . . . any one claiming to have produced such stones is suspect.'
G.F.H. Smith (1940: 138)

The unique properties of the diamond are such that many bizarre hypotheses for its genesis surfaced in the early years following its first discovery in kimberlite matrix. The interpretation by Dr Atherstone was as fanciful as any. In the *Transactions of the South African Geological Society* (1895), he asked how the diamond, which is 'neither soluble nor fusible . . . and is consumed by heat' could survive 'as a crystal in the crater of the volcano'?

Atherstone then claimed that the key to this mystery had been given to him 50 years earlier by the great chemist, Michael Faraday, during a demonstration of the liquefaction of gaseous carbon dioxide by pressure. From this, Atherstone had argued that the Karoo sediments encasing the kimberlite pipes to a depth of at least 2 500 metres would likewise have liquefied this gas generated in carbon-bearing shales. From this stance he postulated that 'the temperature at such great depths being very few degrees above the freezing point . . .we have the constituents of the diamond in a form admitting of crystallization . . . and the subsequent absorption of its oxygen by the iron always present in its containing walls'. This hypothesis would have left this admirable man portrayed in a better light had it never been published! Apart from other objections, the sheer impossibility of his 'proposition' may be gauged from the fact that carbon dioxide requires a temperature of –70 °C under low pressures to form a solid, whereas crustal rocks in fact soon become too hot with depth even to touch with comfort.

Figure 41: A rare specimen of a diamond still embedded in its host rock, kimberlite, in which it normally occurs at a level of a few parts to every million parts of waste rock. This stone is in the region of 7 carats, and its rounded shape is an indication of the resorption (loss of mass, amounting probably to about 50 per cent) that it suffered during the time it was immersed in the hot kimberlite magma during transport from the Earth's mantle towards the surface.

Sir William Crookes (1897b) introduced a novel idea that seduced many of his listeners at the time. He had been deeply impressed by the experiments of the chemist, Moisson, in Paris, who claimed to have produced microscopic diamonds by dissolving pure carbon in molten iron and then quenching the melt in water. Pointing to the known occurrence of diamonds in iron meteorites, Crookes developed the idea that molten iron in the Earth's core was the source of diamonds.

Another popular theme had linked growth of the diamond to reaction between kimberlite magma and carbonaceous shales of the Karoo Sequence. The Ecca Group within the Karoo Sequence is, for example, host to vast deposits of coal in southern Africa, and coal is largely carbon. As diamond mines deepened, this idea was soon discarded, as some pipes showed no evidence of having ever intersected any appropriate carbon-bearing strata (Molengraaff, 1897b).

The South African, H.S. Harger (1905) had had the temerity to reject the theories of the eminent Sir William Crookes. He pointed out that metallic iron had never been found as a natural material in any of the mines. He was far more impressed by a laboratory experiment by a Dr J. Friedlander to the effect that stirring molten olivine with a graphite rod in the laboratory had yielded microscopic diamonds, and that neither 'huge pressure nor great heat' had been necessary. From this shaky base Harger argued for the existence of a carbon-saturated source-zone at depth where 'the richness or poorness in diamonds . . . would be influenced in different places solely by the quantity of the element carbon that was present when crystallization took place'. We may safely assume, however, that Friedlander's conclusion had been based on a misidentification of some other phase as a microscopic diamond (possibly a spinel?). Diamonds do not form in little crucibles at atmospheric pressure.

SPATS BETWEEN GEOLOGISTS AND CHEMISTS

Just short of 100 years ago, F.W. Voit (1907b) highlighted several fundamental questions in the *Transactions of the South African Geological Society*. He was a believer in the crystallisation of diamonds within the kimberlite magma itself, but rejected the idea that carbonaceous sediments had provided the raw material. As proof of this, he pointed out that coal fragments torn from the rocks enclosing the pipes were converted to anthracite, not diamond. In his view, metallic carbides (metals combined with carbon) in the kimberlite magma were responsible. This, however, raised a conundrum – if diamonds were merely transported towards surface from some extensive, deep-seated source-rock, how was the virtually unique character of the stones from each separate mine to be explained?

Molengraaff's arguments had not deterred J.A.H. Armstrong (1907) from resurrecting, once more, the idea of contamination of kimberlite magma by the rocks through which it had passed. He urged exploration of the carbonaceous Ecca Group shales in Natal where they 'have been intruded by granitic rocks of comparatively recent date . . .' This misguided proposal contained at least two fundamental flaws. He had failed to appreciate the massive pressures required for diamond growth, while younger 'intrusive granites' in Natal are poorly represented.

In 1909, W. Johnson, a chemist by profession, had adopted a decidedly supercilious tone in suggesting that 'geologists are sometimes not sufficiently careful in drawing inferences in matters chemical', before launching into his own pet theme. It was hardly a remark that would have endeared him to the community of Earth scientists. Johnson had been impressed by the claim that inclusions of water, carbon dioxide and even a water-bearing mineral, *apophyllite*, had been found trapped within De Beers diamonds, and therefore argued that diamonds could not have formed at temperatures much above 800 to 900 °C – even perhaps as low as 600 °C. In his own words, this 'tells strongly in favour of the diamonds having been formed in the mine [*sic*] after the latter had reached the mud volcano stage'. Put more simply, he opted for diamonds forming near the surface at low temperatures and pressures now known to be orders of magnitude too low. (How the geologists responded to his disparaging remarks about them is not recorded, but in 1914 Johnson did acknowledge that his remarks might have been 'unfortunate'.)

In spite of hundreds of experiments, Johnson was unable to produce synthetic diamonds, but he simply sidestepped this issue by asserting that the duration of his experiments had been too short in relation to the immensity of geological time.

Percy A. Wagner's *Diamond Fields of South Africa* (1914) was a major early milestone. More than 250 occurrences of kimberlite or kimberlite-like rock had by now been uncovered between the Cape province and Katanga in the Congo. The most productive area was that between Jagersfontein in the southern Orange Free State and Premier Mine east of Pretoria, within which 11 of 150 known kimberlite occurrences had been proved economically viable. Wagner, a consulting mining geologist, had enjoyed unrivalled opportunities to examine a great number of these, and was convinced (1914: 131) that '. . . the nodules [xenoliths of deep-seated origin] may be conceived to have resulted from the clotting together under plutonic [deep-seated] conditions of early products of the consolidation of the kimberlite magma, or from the breaking up of bodies thus formed'. He maintained that the diamond had crystallised from this magma, since it was found both as inclusions trapped within typical kimberlite minerals, and as microscopic crystals in the kimberlite groundmass. He did concede, however, that the occurrence of diamonds in the nodules of deep-seated origin (the so-called griquaites) pointed to some additional source. Some specimens of this had by now been found to be spectacularly rich – Corstorphine (1907) had written a short paper on one specimen measuring a mere 12.5 x 7.5 x 4.5 cm from which sixteen diamonds were extracted, and by 1911 he had collected another three similar specimens.

In a further foray in 1914, Johnson, the chemist, revived his theme of diamond crystallisation at low temperatures, now proposing an extraordinary truce to meet certain criticisms by Wagner. This can only be described as an offer of 'don't show me yours, and I won't show you mine'. He offered to drop further discussion about the inclusion within a diamond of the water-bearing mineral, apophyllite, 'just as we shall ask Dr Wagner to leave out of consideration a single case quoted by him where a diamond is alleged to have been found in an unaltered olivine crystal'. Was this a suggestion that they should close their eyes to inconvenient evidence? In any event, from this dubious stance he turned to insist that the presence of oxides of iron, both in kimberlite rock and as inclusions entrapped within the precious stone, was chemically incompatible at high temperatures with the presence of carbon, either as diamond or as graphite. Iron oxides should have been reduced to metallic iron at high temperature, and the carbon of the diamond oxidised to gaseous carbon dioxide or monoxide.

With all the resources of De Beers Consolidated Mines behind him, no one was better placed than its General Manager, Alpheus F. Williams, to take the baton from Wagner and carry it further. This he did in his lengthy two-volume treatise, *The Genesis of the Diamond*, published in 1932. It is fair to say that much of this was a recapitulation and expansion of Wagner's earlier work, but there were also significant differences of opinion. Like Wagner, Williams concluded (1932: 188, 207) that the diamond crystallised from kimberlite magma. In his own words, '. . . it can reasonably be assumed that the bulk of the diamonds did not crystallize until the magma started to rise from its plutonic reservoir' and 'as the magma rose a fall in temperature and pressure followed, with the result that certain minerals started to crystallize in the still fluid magma'. Amongst these 'certain minerals' he included olivine, pyroxenes, mica, garnet and diamond. He dismissed the view of the De Beers Consulting Geologist, Alex du Toit, that the deep-seated nodules were the ultimate source of the diamonds.

DIAMONDS FROM ANCIENT LAVAS?

Not all geologists were convinced during the early 1900s that kimberlite was the sole source of diamonds. F.W. Voit (1907a) was intrigued by the diamond fields in the Bloemhof and Christiana districts, 'spread over a wide area where no alluvial gravel exists' and 'upon the heights where also no traces of an old river gravel is [*sic*] found. Here the diamonds are simply found in the decomposition products of . . . a very green, hard, sometimes amygdaloidal rock of diabasic [lava-like] character . . .' This was the widespread Ventersdorp lava that underlies the Dwyka Tillite of glacial origin, and he made the startling claim that this was the source from which the diamonds were released by weathering.

Surprisingly, Voit's contentious proposition was supported by Dr Hans Merensky (1907), arguably the most successful independent geologist and prospector of the times in South Africa – the man who subsequently amassed great wealth on the diamond fields around the Orange (now Gariep) River mouth, and traced the source of platinum in the Bushveld Complex. It was his contention that two different kinds of diamonds are found in South Africa – 'river stones' possessing superior brilliance, clarity and hardness, and stones mined from kimberlite.

Over an area of some 500 by 80 kilometres that embraced the Klerksdorp, Wolmaransstadt, Bloemhof, Christiana and Barkly West districts, the rock underlying the diamondiferous deposits is generally lava. Diamonds had been concentrated within present or previous river channels, but other deposits were located far from the river channels, without any trace of a kimberlitic source. Merensky and Voit had therefore concluded that the lavas must have been the source of the gems. Their reasoning was, firstly, that as some diamonds are found on the steep sides of *kopjes* (hillocks) they must have been weathered out of the decomposing lava; secondly, that the diamonds were concentrated together with agates weathered out of the lava; and thirdly, that worked-out claims had yielded

Figure 42: The glacial rock, Dwyka Tillite (foreground), resting directly upon polished and ice-scratched Ventersdorp Supergroup lava at Nooitgedacht in the Kimberley district. In earlier days, it was believed by some that the lava was the original source of the alluvial diamonds, which were released as the rock weathered and disintegrated. This interpretation was totally without foundation.

rich new finds when diggers returned to them decades later, following continued decomposition of the lava in the interim.

Merensky's logic was weak, and he had relied too much on the tales and the veracity of the early diggers. The rate at which diabasic rock weathers and decomposes is in reality so low that the assertion that individual claims had in this way been regenerated and restocked with gems within just a few decades was preposterous.

In spite of the obvious objections, this 'theory' was not laid to rest. Harger (1909) added that, over a distance of 400 kilometres between Klerksdorp and Prieska, 'the diamond-bearing gravels . . . always occur in areas where the [volcanic] rocks have been . . . associated with the overlying glacial beds' and that there are diamond sources of different ages – some older than the glacial rocks and others that are younger. And here he dropped what can only be regarded as another red herring. Two tiny pebbles, each less than 3 centimetres in diameter but containing a small diamond, had been sold to him by a digger who claimed that they had come from the Dwyka Tillite. Although they were highly decomposed, Harger insisted that the pebbles were of Ventersdorp lava. This encouraged him to pronounce, with exuberant but misguided confidence, his final conclusion that 'having at last traced the diamonds to their . . . [volcanic] source, it is simple to understand from whence the glacial moraines and gravels of the Vaal River borrowed their precious contents . . . the discovery . . . is a matter of great scientific and commercial importance.' These were stirring words, but the concept was entirely false.

THE IDEAS IN A NUTSHELL

The foregoing paragraphs show that through the early years of the twentieth century the spectrum of 'theories' about the origin of the diamond had been stretched so wide that it had become a matter of personal bias or preference as to what was to be believed.

Even the most untutored digger or labourer on the diamond fields knew that kimberlite was a carrier of the gem. The proof of this was there in abundance in the deep mines. At a higher intellectual level, however, the argument was about whether their ultimate source was in the nodules brought up from great depths beneath the Earth's crust – the olivine-rich peridotites or the eclogites – or whether the diamond had crystallised directly from the cooling kimberlite magma itself.

One popular theme was that the essential carbon had been acquired by contamination by shallow crustal sediments during emplacement of the molten magma. Another attributed the carbon to the presence of metallic carbides in a deep-seated layer, while Sir William Crookes had placed the source even deeper, within the metallic core of the Earth. At the far end of the spectrum of ideas was the claim that the lavas of the Ventersdorp 'System' were a source. As for the physical conditions under which the diamond was formed, estimates of temperature ranged from 600 °C upwards, and pressures from atmospheric to those prevailing in the Earth's core. The 'theories' were as diverse as the prejudices of those who proposed them.

DIAMONDS SPREAD BY ICE SHEETS?

Harger's contention that the Dwyka Tillite was a carrier of diamonds may be re-evaluated in terms of a recent review of the origin of the west coast deposits of southern Africa by Moore and Moore (2004). They have stressed that 'the source area of Dwyka Group glacial sedimentary rocks in southern Africa contains a province of pre-Karoo diamondiferous kimberlites', and maintain that diamonds and kimberlite indicator minerals had been scraped off the surface of this source region during the Carboniferous glaciation, then transported as the ice sheets moved towards the west, and, finally, released by subsequent disintegration of the host Dwyka Group rocks.

Moore and Moore quote undeniable evidence for the transportation of rock fragments and boulders for distances of 900 to 1 000 kilometres by the ice sheets, which brings into focus the possibility that diamonds from the region of the Premier pipe east of Pretoria were also carried south-westward. The feasibility of the concept seems to be borne out by similar patterns of dispersal of the indicator minerals, and the diamonds accompanying them, that have been mapped in Canada and

Brazil. This elegant synthesis suffers, however, from one severe constraint – repeated sampling of the Dwyka Tillite by mining companies has over several decades failed to show that the gems actually occur within the ancient glacial sediments. A final answer to this controversy is still to be found.

* The diamond has always been regarded as the hardest naturally occurring substance known, and graphite as amongst the softest. Huntress (2003), however, has noted that chemical vapour deposition has produced artificial diamond that is harder still than natural diamond. Furthermore, a synthetic form of graphite has been produced, which is 'superhard' to the degree that it is capable of indenting diamond.

PHYSICAL CHEMISTRY CRACKS THE PROBLEM

The years of controversy had wrapped an aura of mystery around the diamond. What was the true history of this unique stone? It was from the laboratory that the answers eventually came. The diamond is simply crystallised carbon. However, so too is graphite, a soft mineral best suited to the making of pencil 'leads', black polishes and lubricants. The difference between the two is merely the difference in the crystalline structure adopted, that is, the way in which the carbon atoms are chemically bonded together in the crystals.* Within the range 1 100 to 1 600 °C, the one form will invert to the other if the pressure changes, diamond forming at higher pressures, and graphite at lower pressures.

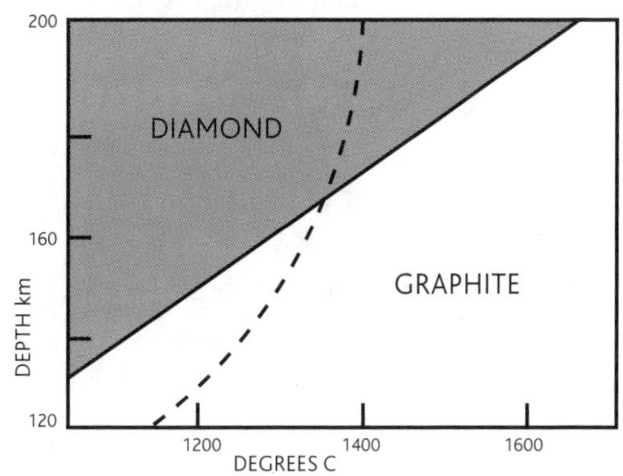

Figure 43: The relationship between diamond and graphite. Diamond is stable only within the high-pressure (deeper) region indicated by the shaded area of the diagram. At lower pressures graphite is the stable form. If, however, the pressure is lowered sufficiently rapidly (diamond rapidly transported to surface), the inversion to graphite may fail to take place, and the diamond is permanently preserved in the metastable form. The broken line shows the natural increase of temperature and pressure deduced to prevail in the Earth's interior, indicating that diamonds probably do not form at depths less than around 150 kilometres.

If, however, the diamond is rapidly brought down to room temperatures and pressures under closely controlled conditions, it will bypass the inversion and retain the high-pressure structure and fail in the time available to become reorganised into the graphite form. It is then permanently preserved in what mineralogists term the metastable state – a 'freezing' of the diamond structure down to low temperatures. This is the form in which all diamonds occur on the surface of the Earth. If the diamond did not 'miss the bus' like this, diamonds would be unknown outside the laboratory. Diamond rings, tiaras and bracelets would be mere figments of the physical chemists' imagination.

All of this has profound implications in unravelling the genesis of the diamond. The pressures appropriate to the crystallisation of carbon as diamond, rather than as graphite, are reached at a depth of not less than 150 kilometres below the Earth's surface. At shallower depths, graphite would be the stable form.

It is now known that, once formed, the diamond crystals may be stored for immensely long periods below the Earth's crust, before being ejected towards surface. Dating of the mineral inclusions that grew at the same time as the diamond itself (Richardson, 1986; Richardson et al., 1984, 1990; Kirkley et al., 1991) has shown that this time lapse may be as great as 3 200 million years. Although some diamonds may have been formed around one-third of the time that has lapsed since the Earth itself was formed, the emplacement of them near the surface may be a relatively recent event that took place less than 100 million years ago. At Finsch Mine, the diamonds are of different ages, depending on whether they are associated with minerals of the peridotitic nodules (3 200 million years) or eclogitic nodules (1 580 million years). This discovery established that the kimberlite magma was little more than a liquid medium that had sampled diamond-bearing source rocks at great depth, and then carried the stones towards the surface. The kimberlite magma had merely been an agent of transport, like a lift in a skyscraper. With the discovery of diamonds in Australia, it has also emerged that kimberlite is not the only possible transporting medium. Here, potassium-rich rocks known as *lamproites* are the carriers. This raises the intriguing possibility that yet other, unsuspected igneous rocks could conceivably be diamondiferous, provided that they originated at very great depths below the surface.

An important requirement – one that in part explains the relative rarity of diamonds – is that its transportation to surface must be achieved with speed in order to prevent it reverting to graphite. The speeds attained by the transporting medium have been estimated by some to be 10 to 30 kilometres per hour in the initial stages, reaching perhaps several hundreds of kilometres per hour as the surface is approached (Kirkley et al., 1991). Exceptional velocities of this order are the result of the kimberlite magma being ejected from a region of high pressure, and being highly charged with compressed gases and volatiles – a property not shared with many other magmas.

The inversion of diamond back to graphite has been recognised in some deposits, and is analogous to the occurrence of minute cubes and octahedral crystals of graphite (showing the shapes of diamond crystals) found in meteorites, where they are referred to as the mineral species *cliftonite* (Ford, 1945: 397). Furthermore, if the transportation to surface is slow, partial solution of the diamond in the kimberlite magma, or its conversion to a gas may take place. For this reason, many natural diamonds are somewhat rounded, with their sharp edges blunted by corrosion (see Figure 41).

DIAMONDS AND PLATE TECTONICS

Armed with the ability to distinguish between diamonds from eclogites (those termed E-type) and those from peridotites (P-types), diamond, scientists turned to examine the nature of the carbon of which the diamond is made (see Kirkley et al., 1991). In nature, carbon exists as different atomic species known as carbon-12, carbon-13 and carbon-14. Measurements of variations in the ratio between the isotopes carbon-13 and carbon-12 have revealed characteristic differences between E-type and P-type diamonds. Those of P-type show very little variation in the range of these isotopic ratios, which suggests to geochemists that the source rocks reflect the type of carbon that was an original component of the primitive Earth at the time of its formation.

The E-type diamonds, however, show wide variations in the ratio – a feature generally characteristic of shallow crustal rocks. This brings into focus the Plate Tectonic Model (see Chapter 6). It may be conjectured that when drifting continents collide on the surface of the planet, wedges of rock are thrust (subducted) deep into the mantle beneath the crust, where the extreme pressures and elevated temperatures form E-type diamonds.

THE FINAL DISPELLING OF A MYTH

The curious thing about diamonds is how few theorists came even close, prior to the 1950s, to the truth about the gem's origin. The aura of mystery and romance that had wrapped itself around the stone through several millennia nurtured a belief that this was a unique substance, and that no tricks of science could ever be used to produce the stone artificially. The assertion by Herbert Smith, quoted at the start of this chapter, that: '. . . any one claiming to have produced such [artificial] stones is suspect . . .' was the prevailing dogma. But, like so many old myths, this one has been laid to rest. Less than twenty years after Smith's bold assertion, synthesis of the diamond became a reality.

Natural diamonds fall into two categories – gemstones and the flawed or smaller stones used for industrial purposes. Their natural hardness renders them invaluable for manufacturing diamond saws, cutting tools, diamond drills, engraving tools, etc., and for the preparation of diamond pastes and suspensions for polishing hard materials, including other diamonds. These are as indispensable in industry as are electricity and oil-based fuels in the modern world.

It was the fear that the mines of the world would not be able to meet the demand for industrial stones in post-Second World War times that spurred research into the possibility of synthesising them. In 1955, General Electric in the USA claimed that it had been successful in this, but, in order to keep its process secret, it was not patented until 1959. In the meantime, the De Beers Group had in 1958 quite independently produced its first diamond, and in 1960 Ultra High Pressure Units was formed in Springs, South Africa. From there operations expanded with the opening of new plants in Ireland.

By the 1970s, the production of synthesised industrial diamonds exceeded the sales from the mines, and by 1990, 85 per cent of the 160 million to 180 million carats of industrial diamonds consumed each year were produced by synthesis (Wannenburgh & Johnson, 1990: 121). Armed with the knowledge that earlier pages have summarised, the reader will have guessed that the plant for producing the synthetic stones employs high temperatures within 10 000-tonne presses for exerting enormous pressure – as in the Earth's mantle.

The last bastion that had preserved the mysticism surrounding the diamond had been breached. The ancient cloak of romanticism was stripped away, and those whom Wannenburgh and Johnson christened 'the New Alchemists' had won the day.

32
DIAMOND PIPES: NATURE'S DRILLHOLES THROUGH THE EARTH'S CRUST

'How the great pipes were originally formed is hard to say. They were certainly not burst through in the ordinary manner of volcanic eruption . . .'

Sir William Crookes, FRS (1897a), in a celebrated lecture

'The pipes represent deeply eroded, funnel-shaped volcanic necks . . . The earliest eruptions appear to have been in the nature of mighty gaseous explosions, which resulted in the blowing-out of funnel-shaped apertures.'

Percy A. Wagner (1914: 5, 41)

The Kimberley 'Big Hole' (see Plate 19) probably ranks as one of the most photogenic man-made features in all of southern Africa. Even today, more than 130 years after mining started there, its visual impact stirs the imagination and, for many, induces an attack of vertigo. Standing close to the rim, you look down upon a great, roughly oval hole plunging into hundreds of feet of stagnant green water ponded at the bottom. The dimensions of this excavation in 1932 were recorded by Alpheus Williams as 38 acres (15.4 ha) at surface, with a rim-to-rim diameter of some 1 500 feet (460 m), and the depth from the level of the surrounding flat plain to the surface of the pool as close on 1 000 feet (300 m).

Although mining by opencasting from surface had long ago been discontinued at a depth of 400 feet (120 m) because of the hazard of falling rock, the depth of the hole open to the sky had eventually reached 1 300 feet (around 400 m) as the diamond-bearing rock was progressively drawn out from below, along deep haulages connecting with a vertical shaft reaching 3 600 feet (1 097 m) in depth. Collapse of a collar of soft and easily weathered shales, representing the uppermost 100 metres of older strata rimming the pipe, has through the years produced a funnel-like flaring of the top of the great pit, giving a false and exaggerated impression of the dimensions of the pipe itself. Below this upper funnel, smooth walls of hard lava and quartzites that encased the diamond-bearing core show the true dimensions of the pipe proper – still more than 7.3 hectares in area at a depth of 300 metres (declining eventually to 2.5 ha at 1 066 m). Now there is only silence where thousands of men once toiled, while swallows circle through the air far below, and pigeons nest on the rocky ledges.

QUESTIONS . . . AND MORE QUESTIONS . . .

Despite the impact that the Kimberley pipe makes on the imagination, it is in reality one of the smaller pipes – about one-third of the size of that at Koffyfontein, and one-eighth that of the Premier Mine. Whether you are a geologist or not, questions crowd in on the mind. How were these carrot-shaped holes drilled through the immensely hard rock of the Earth's crust? Was this once a scene of violently explosive activity, or did the kimberlite insinuate itself slowly and tranquilly, eating away the overlying column of rock? Since the undisturbed attitude of the surrounding strata shows no signs of their having being pushed aside to make room for the infilling, where is the spoil that must have been thrown out to create the massive void into which the kimberlite was emplaced? How far above the present level of exposure did the pipe, or others like it, extend before erosion stripped off overlying strata? Did the pipe-like conduits reach the surface at one time to form cone-shaped volcanoes?

. . . AND HOW THEY WERE ANSWERED

There is some argument about whether it was the government geologist, E.J. Dunn in 1871, or the visiting German geologist, Professor Cohen in 1872, who first recognised that the drilling of the pipe-like conduits was related to some form of volcanic activity. The evidence was incomplete, since the uppermost parts of the pipes where they might originally have burst through to the surface had long since been eroded away. Thousands of metres of lavas and fossil-bearing strata that the tops of the pipes had penetrated had also disappeared. There was no way of determining accurately how much was missing. Some contemporary estimates had speculated that this might have been as much as 6 kilometres, but today we would dismiss this figure as excessive. Whatever the figure might have been, the exposed pipes obviously represented only the deeper parts of the original structures.

Never at a loss for words on any topic, Ernest Schwarz (1914a) had tossed into the melting pot of ideas his opinion that, 'the great fault that I find with people who argue about the diamond pipes . . . [is that they forget] there was once a vast superstructure which is similar to that of other volcanoes'. Apart from the use of the word 'vast' he was not far off the mark, but his listeners would surely have had severe misgivings about his expertise in this field after his controversial claims (Schwarz, 1908) that 'fresh water mussels, blocks of wood and even an assegai head were found down to 150 feet'.

The interpretation reached in 1914 by the leading authority of the times, Dr Percy A. Wagner, was that the ascent of kimberlite magma from great depth was in the initial stages guided by structural weaknesses in the Earth's crust, and that it had been injected quietly as dyke-like bodies – the so-called fissures (see Plate 20), many of which are richly diamondiferous and still being mined today. Only after this, he argued, were the pipes drilled by 'mighty gaseous explosions which resulted in the blowing out of funnel-shaped apertures' (1914: 41). The bulk of the

Figure 44: Koffyfontein Mine, reopened in the 1970s, showing benches cut to facilitate opencast mining of the kimberlite pipe towards the left.

material that had been fragmented or blasted out then fell back into the vents, as evidenced by the 'rubble' or breccia that is today found within them. In his view, it was only in the final stages that the kimberlite magma rose to fill the pipes during a succession of eruptions accompanied by violent churning and comminution of the contents. In this way was the load of exotic fragments – the xenoliths – thoroughly intermixed to bring fragments of both deep-seated and shallow provenance into juxtaposition.

Where Wagner (1914: 41) had distilled his final conclusions to fit into little more than a single page of text, Alpheus Williams (1932) spread his analysis through more than 100 somewhat discursive pages in his two-volume treatise, *The Genesis of the Diamond*. Here he contested virtually every tenet put forward earlier by Wagner. He interpreted the pipes as mere enlargements along fissures, and argued that the kimberlite had been emplaced 'without undue violence through fissures'.

The stalemate was broken only when unexpected evidence of a different sort turned up – that yielded by pipes in other countries where only minimal erosion had occurred since the time they broke the original surface. Prospecting in the bush of Tanzania, a Canadian geologist, John T. Williamson (a graduate of McGill University) had discovered the Mwadui pipe in 1940, some 160 kilometres south of Lake Victoria. What was exposed here was startlingly different to the pipes known at that time in southern Africa. Beneath superficial deposits of soil and diamondiferous gravels was a basin of khaki-coloured shales capping and dipping inwards towards the centre of a very large pipe. The structure of the basin, and the fossil remains of insects and frogs, clearly identified the shales as fine-grained sands and muds that had accumulated in a crater at the top of a pipe. Drillholes that were put down in the 1950s showed that the shales were at least 200 metres thick, and that with increasing depth there was a gradual transition to kimberlitic debris that had been laid down by sedimentary processes.

The separate pieces of the puzzle were now beginning to fall into place. An active research team had been growing under the leadership of J. Barry Hawthorne, Consulting Geologist within the De Beers Group, and a unit devoted to fundamental research was created at the University of Cape Town under Professor John Gurney. The First International Kimberlite Conference was launched in Cape Town in 1973, and was followed by further meetings every few years. In 1975, Hawthorne put together a model that is, essentially, only little changed today, but which tied everything together.

THE HAWTHORNE MODEL

This model (Hawthorne, 1975) recognised the passage of deep feeder dykes of kimberlite into the base of the pipe-like structures in a *root zone* within some 3 kilometres of the surface. At this depth, the kimberlite displays the features of typically igneous rock. Above this, within a pipe that expands upwards, lies kimberlite which shows the attributes of a fragmental rock (the *diatreme zone*) produced by the explosive release of pent-up gases, largely carbon dioxide and water. This is the typical filling of the South African pipes that were exhumed by the early miners in the Kimberley region, and from which the great bulk of our diamonds came.

Within this diatreme zone, the expansion of the released gases would have caused rapid cooling of the emplaced material, thus accounting for the absence of any indications of high temperatures. The uppermost few hundred metres, typified by the exposures at Mwadui, lie within what Hawthorne defined as the *crater zone*. This was where the surface expression was essentially that of a volcano, but not of the classic type of cone-shaped edifice built to impressive heights by repeated and prolonged extrusion of lavas, such as can be seen at Mt. Taranaki in New Zealand, or Mt. Fuji in Japan. Merely a low ring of fragmental material would have encircled a central crater. Gradual infilling of this depression by transport and slumping of the debris back into it would have imparted a bedded character to the uppermost deposits. Eventually, these infillings would have graded upwards into shallow-water sediments typified by the shaly cap.

Geologists are an argumentative bunch, and the focus of controversy has now shifted to question what role underground water might have played in generating the massive explosions that drilled the pipes. Some now argue that water coming into contact with hot magma would have flashed explosively to steam. Comparisons with maar-type volcanoes in Europe, which are indeed created by such a process, sustain current arguments about whether it was the release of compressed gases within the kimberlite itself, or its contact with underground water, that was the cause of the explosions.

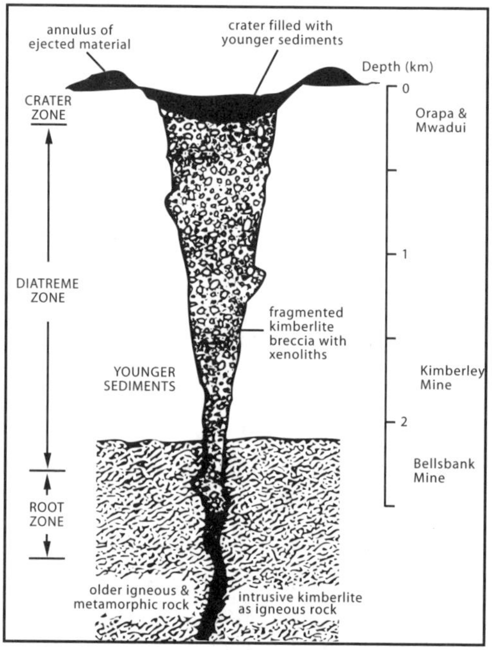

Figure 45: Idealised section through a kimberlite pipe, as modelled by J. Barry Hawthorne (1975). The most productive zone is the ***diatreme zone***, as at Kimberley. The deeper ***root zone*** where a dyke-like form is adopted may be diamondiferous, but the available tonnages of ore are low. The area over the sediment-filled ***crater zone*** may develop rich, shallow deposits due to erosion of the land surface, leaving diamondiferous gravels or conglomerates, as at Mwadui Mine in Tanzania.

Figure 46: Mwadui Mine, Tanzania, in the 1950s. The efficiency with which nature had hidden this pipe is shown by the thick layer of 'black cotton soil' blanketing surface quartzite (in the process of being stripped off and dumped to the left of the dragline). The underlying diamondiferous gravels (on which the ore carrier is parked) are revealed in the middle foreground. Where 10 to 15 carats of diamond per 100 tons of ore is, in general, a workable grade in diamond mines, these gravels yielded 600 to 800 carats/100 tons in the 1950s. This mine yielded a 56-carat rose-coloured diamond that was presented to Princess Margaret Rose.

33
THE PROPERTIES THAT MADE GOLD AND SILVER VALUABLE

'If gold ruste, what shall iren do?'
Geoffrey Chaucer (1340–1400), in The Canterbury Tales *(Cohen & Cohen, 1963: 107–14)*

'Gold! Gold! Gold! Gold!
Bright and yellow, hard and cold.'
Thomas Hood (1798–1845), in Bartlett (undated: 292)

'From the red gold keep thy finger.'
Walter Scott (1771–1832), 'The Bride of Lammermoor' (Cohen & Cohen, 1963: 306–25)

'To gild refined gold, to paint the lily,
Is wasteful and ridiculous excess.'
*William Shakespeare (*King John*: IV, ii, 11)*

'We will answer their demand for a gold standard by saying . . . you shall not crucify mankind upon a cross of gold.'
William J. Bryan, National Convention, Chicago, 1896 (Browning, 1951: 38–428)

'All is not gold that glitters [or glisters].'
Ancient Latin proverb (Cohen & Cohen, 1963: 178)

The grip that gold has long held on the human imagination is implicit in the abundant references to its properties in numerous compilations of familiar quotations culled from our literature. It is an extraordinary fact that virtually all civilisations have been obsessed by the metal from the earliest times of recorded history. De Kock (1994) has speculated that as far back as 70 000 years ago, Neanderthals might have collected gold nuggets to string together as ornaments, while the earliest known wrought gold jewellery dates from the Sumerians around 3000 BC. According to Van Wyk (1995) the first gold coinage dates back to about 700 BC in Asia Minor, when the natural gold-silver alloy, electrum, was stamped with engraved punches. The legendary King Croesus subsequently introduced separate gold and silver coinages to avoid the disadvantage of variable composition of the alloys inherent in the use of electrum. From here the use of gold coinage spread to Greece and Rome.

There is scarcely a tribe in history that did not prize the yellow metal and use it for ornamentation, coinage or trade. It figured prominently in the cultures of the Sumerians, Egyptians, Phoenicians, Indians, Arabs, Greeks and Romans, Incas, Spanish, Portuguese and English, and in later days in the mining industries of southern, Central and West Africa, the Americas, Australia and New Zealand.

A BLOOD-SOAKED HISTORY

The translation from the Latin of Agricola's (1556) *De Re Metallica* by Hoover and Hoover (1912) offers a vivid picture of mining in ancient times. Their 'Historical Note on Crushing and Concentration of Ores' (1912: 279–83) notes that the earliest indications of these activities appear on monuments of the IVth Dynasty of Egypt (4000 BC) and stelae of the XIIth Dynasty (2400 BC). A horrifying account of the practices in Egypt, Arabia and Ethiopia, penned in the second century BC by the geographer, Agatharchides, describes mining by 'a vast multitude of people' composed of criminals, captives of war, or those who had merely offended the king in one way or another, together with their women, children and other relatives, condemned to work naked and 'all bound in fetters'. These unhappy wretches were supervised by brutal overseers who 'lashed them severely' while forcing them to labour in the darkness of the excavations, with lamps on their foreheads, until 'at length, overborne with the intolerable weight of their misery, they drop down dead in the midst of their insufferable labours'.

The scale of operations in the silver mines at Mt. Laurion, where production peaked before the third century BC on the southern end of the Attic Peninsula of Greece, may be gauged from the reworking of ancient waste dumps in the nineteenth century: some of these dumps were estimated at 7 million tons. An output of approximately 84 000 troy ounces was claimed for the year 484 BC (1 troy ounce = 31.1 grams). An account of mining in Spain (New Carthage) by Polybius (200–125 BC) refers to a workforce of 40 000 men yielding a daily revenue of 25 000 drachmas,* and Spanish silver mines at Baebelo worked by Hannibal (247–183 BC) yielding 140 kilograms weight of silver per day (Hoover & Hoover, 1912: 42, 280).

* According to Funk (1903) the ancient drachma was a silver coin with a value variously defined as equivalent to US$0.09 to US$0.6. Using the higher figure, 25 000 drachmas would equate to some 15 000 nineteenth-century US dollars.

Historians tell us that the mighty Inca Empire of South America amassed huge quantities of gold and silver long before the arrival of the Spanish conquistadores in 1532. This was not used for trade or currency, but in jewellery and objects of religious significance. Pizarro, it is told, demanded that one large room be filled with gold objects to the height of his raised arm, and another with silver, before he treacherously murdered the Inca Emperor, Atahuallpa, in 1533.

THE PROPERTIES OF GOLD AND SILVER

What was it that drove one civilisation after another to seek these metals so avidly? Even communities that were geographically remote and isolated from one another, and which could not therefore have come to acquire their perceptions from others, valued them highly. Common sense tells us that this must be attributed to the sum of the properties of gold, and to a lesser extent those of silver.

Gold is a bright metal with a warm, sensuous colour because it is a more efficient reflector of light at the red end of the visible spectrum than for blue. Actual values (Eales, 1967) range from a reflectivity better than 90 per cent for light at a wavelength of 700 nanometres (red light) to less than 30 per cent at 400 nanometres (deep blue).

In its natural state gold is never truly pure, as it is alloyed with variable amounts of silver, or small amounts of base metals such as copper, nickel, etc. Depending on its composition, its colour ranges from the reddish hue of the pure metal to a pale-yellow alloy (electrum)

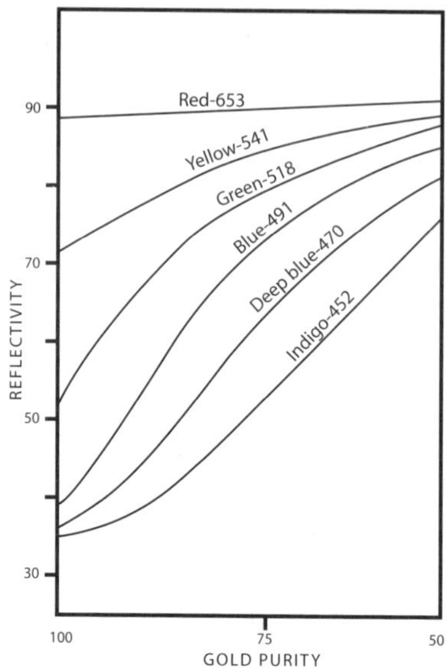

Figure 47: Measurements of the reflectivity (brightness) of gold-silver alloys in visible light of different colours (wavelengths given in nanometres [nm]). The composition of the alloys is shown (as the percentage gold) on the horizontal axis of the diagram, and the reflectivity on the vertical axis. For all compositions the reflectivity decreases markedly as the incident light shifts towards the blue end of the spectrum, while platinum and silver retain their brilliance. Even the common, virtually valueless mineral, pyrite (FeS_2 or 'fool's gold') has greater brilliance in blue light (at wavelength 452 nm: 44.7 per cent for pyrite, as against 35.6 per cent for pure gold), and the precious metal appears dull and unattractive (Eales, 1967).

as the silver content increases. This has been understood since ancient times, when the touchstone was used to test its purity.*

Gold does not dissolve in the common acids – nitric, hydrochloric or sulphuric – or tarnish on exposure to air. Even the golden mask of Tutankhamen taken by Carter from the tomb within the Valley of the Kings in Egypt is as bright today as when it was crafted those thousands of years ago.

Historically, silver and copper have been valued below gold: this is surely due to their tarnishing rapidly. A property that would have been a boon to the early goldsmiths is its low melting point (1 063 °C, declining to 960 °C as the silver content increases), which can readily be attained in a charcoal furnace. Hence it is easily smelted and moulded using even primitive technology. Its high density (19.3 gm/cc – nearly twice that of lead) gives it a feeling of solid worth. It is also amongst the most highly malleable and ductile of all metals, which means it can be beaten into the thinnest of sheets of gold leaf, or drawn into wire. Finally, it is rare in its host rocks, averaging some five to ten parts to every million parts of waste in an average ore deposit, so it is not won without strenuous effort.

* The touchstone was a fine-grained, black rock with a flat, matte surface, across which the gold sample was drawn so as to leave behind a line (the 'streak'). The colour of the line was then compared with those left by standards of known composition. Healy (1986: 126) tells us that the Roman, Pliny the Elder, was enthusiastic in endorsing the accuracy of the test, but completely misunderstood its use in claiming that it yields an assay of the ore itself, which is a quite different thing.

Medieval alchemy was obsessed by the belief that base metals could be turned into gold or silver. Agricola leaves us in no doubt of his derision in his book, *De Re Metallica*, (1556: xxviii):

> these alchemists, since they deceive people, are not only held in the greatest odium, but their frauds are a capital offence . . . we do not read of any of them ever having become rich . . . [although they] are striving night and day to that end . . . There is another class of alchemists who do not change the substance of base metals, but colour them to represent gold or silver . . . A third sort of alchemists . . . throw into a crucible a small piece of gold or silver hidden in a coal, and after mixing therewith fluxes . . . pretend to be making gold from orpiment, or silver from tin . . .

This did not dissuade Agricola from listing a number of prominent names who pursued the 'art of alchemy', amongst them Merlin the Magician, and Cleopatra of Egypt.

34
THE GOLD MINES OF MONOMOTAPA: WAS THIS THE BIBLICAL OPHIR?

'I [King David] have given to the house of my God . . . even three thousand talents of gold, of the gold of Ophir, and seven thousand talents of refined silver . . .'

I Chronicles, 29, 3–4

'And Hiram sent in the navy his servants, shipmen that had knowledge of the sea, with the servants of Solomon. And they came to Ophir, and fetched from thence gold, four hundred and twenty talents . . . And when the queen of Sheba heard of the fame of Solomon . . . she came to Jerusalem with a very great train . . . and very much gold, and precious stones . . . and she gave the king an hundred and twenty talents of gold . . . Now the weight of gold that came to Solomon in one year was six hundred three-score and six talents of gold . . . and king Solomon made two hundred targets of beaten gold . . . and three hundred shields of beaten gold; three pounds of gold went to one shield . . . and all king Solomon's drinking vessels were of gold, and all the vessels of the house . . . So king Solomon exceeded all the kings of the earth for riches . . .'

I Kings, 10–11

'I . . . discovered the first goldfield on the 3rd September and, on the 5th September, the ruins of Zimbabye . . . Having read much about the presumed site of Ophir, the region between the Limpopo and Zambesi appears to me as the most likely one . . . I believe I do not err when I suppose that the [Zimbabwe] ruin on the mountain is an imitation of the Solomonic Temple on Mount Moria, the ruin on the plain a copy of that palace in which the Queen of Sheba dwelled on her visit to Solomon.'

Karl Mauch (1871), explorer and geologist (Bernhard, 1971: 233)

Were the Zimbabwe gold mines those of the Queen of Sheba, as claimed by Karl Mauch, geologist and explorer, and by Carl Peters (1902), the would-be builder of a German empire in Africa, as well as many others who believed the same? Or was Ophir in India, or perhaps Arabia, as still others supposed?

Historical accounts leave us in no doubt that significant amounts of gold were mined in Egypt, Europe, Arabia and Ethiopia in ancient times, but the location of Ophir remains shrouded in mystery. We have no reason to believe that the poet, Milton, was any better informed than others when in 1674 he penned in his *Paradise Lost* the words:

> Mombasa, and Quiloa, and Melind,
> And Sofala thought Ophir, to the Realme
> of Congo, and Angola fardest South . . . (Book XI, lines 399–401).

That Ophir was certainly a rich source of the metal is illustrated by the substantial addition to King Solomon's wealth through a period of just one year, as recorded in the biblical references at the start of this chapter. Using a conversion of one talent as 837 troy ounces (Hoover & Hoover, 1912: 29), the addition to his coffers of 666 talents would have amounted to some 560 000 troy ounces. In the currency of our present times, the value of this would be US$200 million to US$300 million, or R1.5 billion to R2.2 billion; and the weight of gold from Ophir that King David contributed to the building of the Temple would have been considerably more than four times that amount. Nor is the gift of 120 talents to the coffers of Solomon from the Queen of Sheba to be regarded as trivial.[6]

The extravagant use of gold in decorating and furnishing the Temple, and in the crafting of utensils, is astonishing – if the ancient inventories are to be believed. Quoting from such sources in his history of Mount Moriah in Jerusalem, *Blood on the Mountain*, Andrews (1999: 48) paints a picture of the wealth of Solomon. Every visible surface within the Temple, including the ceilings, was coated with gold leaf, while his craftsmen were ordered to manufacture:

- 10 000 golden tables;
- 80 000 gold and 160 000 silver dishes;
- 20 000 gold and 40 000 silver vials and cups;
- 60 000 gold and 120 000 silver basins;

- 10 000 gold candlesticks;
- 40 000 musical instruments, including harps of electrum; and
- 70 000 gold censers for burning incense.

SIXTEENTH-CENTURY ZIMBABWE – AND LATER

Mauch's claim to have 'discovered' the Zimbabwe Ruins, and that, '. . . on 5th September, 1871, I was lucky enough to be the first white man to set eyes upon them' (Bernhard, 1971: 233) may have been legitimate, but could also be interpreted as an extravagant bid for fame in keeping with the flamboyant character of the man – yet at odds with his own acknowledgement that 'Zimbaoe' was known from early Portuguese writings. In his paper, 'The Less Known Ruins in Rhodesia', Franklin White (1903b) draws attention to references made by early Portuguese explorers to the stone settlement now known as Great Zimbabwe yielding in the sixteenth century 'objects of a mysterious origin'. Amongst items found during subsequent archaeological investigations, White lists iron and brass cannons, crockery, glass and beads attributed to occupation by the Portuguese.

Figure 48: Ninth-century stone carving depicting the wisdom of the great King Solomon. Here he is shown judging the ownership of the child claimed by two different women, as told in I Kings 4 of the Christian Bible. This frieze can be found on the west wall of St Declan's Church, Ardmore, County Cork, Ireland.

The writings of the historian, De Barros, and by Father Montclaro in 1572 are shown by Mennell (1904) to leave little doubt that the settlement was indeed known to explorers long before the time of Mauch. Similarly, in *The Empire of Monomotapa* translated from the French edition, Randles (1975) includes extracts from letters written by Diogo de Alcáçora in 1506 and Gaspar Veloso in 1512, from which it is clear that Portuguese traders had by that time already penetrated far inland; by 1552, according to De Barros, they had explored well within the borders of present-day Zimbabwe. From the earliest times, the interior was referred to as Monomotapa – the title also of the hereditary ruler.

There is also an intriguing description of ruins, unmistakably those of the Great Zimbabwe site, from the pen of De Barros, although whether this description was passed on by Arab rather than Portuguese traders seems uncertain. These written accounts would, nevertheless, seem to confirm that the ruins were known to early adventurers long before the time of Mauch.

It is probable (Randles, 1975: 3) that the fortress of Great Zimbabwe had already been abandoned and the focus shifted to Khami (near present-day Bulawayo) by the early sixteenth century, when the Portuguese penetrated the interior from Sofala, in what is today coastal Mozambique. The Rhodesian archaeologist, Summers (1969: 206–08), attributes its decline to both climatic and geological factors – progressive degradation and silting of the Sabi River until it was no longer a navigable trade artery. The Portuguese occupation of Sofala proved, however, to be an unsuccessful venture, as the focus of long-established trade between the indigenous peoples and Arab merchants then shifted further north, and relations with the Monomotapa, ruler of the kingdom, became increasingly uneasy over the years.

Portuguese expansionism had reached its peak by 1652, and in 1683 the Monomotapa closed all the mines under his control. Finally, in 1693, an uprising virtually terminated the Portuguese presence on the high interior plateau, and trade far inland effectively came to an end. From records compiled by Randles, the export of gold through Sofala declined from 5 500 to 8 500 kilograms (177 000 to 273 000 troy ounces) per annum before the beginning of Portuguese occupation, to around 1 500 to 850 kilograms in the seventeenth century, and 1 000 to 500 kilograms in the eighteenth. By 1820, this had dwindled to a mere 11 kilograms.

As early as 1512, Veloso had referred to a further kingdom that lay deeper within the interior than Monomotapa, ruled by the King of Butua from its capital at Khami and possessed of 'great reserves of gold'. The subsequent denial of access to this region was conceivably a contributory cause of the decline of Sofala as a centre of trade.

THE SMALLWORKERS OF RHODESIA (ZIMBABWE)

When the first white settlers arrived in Zimbabwe (then Southern Rhodesia) late in the nineteenth century, they found a country pockmarked by some 4 000 pits, trenches and shallow shafts from which gold ore had been mined, and another 500 sunk on copper deposits (Summers, 1969: 107). There was no written record of who these early miners were, or how they had acquired their skills – they were simply referred to as 'the Ancients' by the settlers.

Time has taken its toll on the ancient mines. Many survive as no more than shallow depressions filled by earth, leaves and debris that have washed in through the centuries. Others still preserve cave-like openings or quarries driven into the hillsides, and a few are still accessible to significant depths. It was these vestiges of prehistoric activity that had guided those amongst the wave of new immigrants from the West who had come to prospect and mine in the newly opened territory. For, whoever they were, 'the Ancients' had been skilled and efficient prospectors, and hindered in their pursuit of metals only by a lack of adequate tools and technology.

A well-known South African geologist, Dr Gerry Söhnge (1946: 14), has commented on ancient copper workings in the Messina district, south of the Limpopo River, and observed that not a single ore body exploited in modern times by Messina Copper Mines was without evidence of having first been worked by 'the Ancients'. All but a small proportion of the gold mines established in Southern Rhodesia were developed on pre-existing workings. In this way an immigrant community who came to be known as 'smallworkers' grew up. No hardier, ever-optimistic or more innovative breed was to be found anywhere in the modern world.

The dusty files stored in the offices of mining commissioners in the bigger towns of the territory tell us something of these men, where they came from, and what success they achieved. Some names given to claims pegged hint at the independent, pioneering spirit that drove them to search in remote places, or suggest what they found in the ancient workings they discovered there: Antbear, Edge o' Beyond, Lonely, Lone Hand, Black Mamba, Owl's Nest and Bushtick claims. Other names give clues to the places they hailed from: Geelong Mine, New Canada, Emerald Isle, Ben Nevis, Big Ben, Caberfeidh, Cambrian and Penzance. Evocative of their eternal optimism were names such as the Bye-and-Bye, Champion, Dan's Luck, Gee Whiz, Little Beauty, Eldorado, Just-in-Time and Never Despair claims, while final acceptance of repeated failures is recorded in the Fed Up, Last Chance and R.I.P. claims. Records of the gold they produced from individual claims ranged from a recovery of less than one ounce to others where fortunes were made on mines that survived for scores of years.

The deposits of this territory are in no sense to be compared with those of the Witwatersrand, the largest gold deposit in the world, where ancient sediments were long ago washed into a shallow sea.

With the sediments had come the gold particles released by prolonged erosion of a hinterland that had held countless thousands of smaller veins, lodes and impregnations – rather like Zimbabwe today where the smallworkers mined. Some smallworkers crushed the veins of white quartz and caught the gold particles that were released on mats of corduroy or copper plates covered with mercury. The crushing was done by stamp mills powered by suction-gas engines that ran on green firewood. Others extracted gold by direct cyanidation of ore from wider zones impregnated by the passage of mineralising fluids. Here they dug out the soft, oxidised and rotted rock of the first 15 to 20 metres below surface, and threw it into ponds or tanks filled with a weak solution of sodium cyanide to dissolve the gold particles. When the shallow ore ran out, they simply moved on.

To encourage production, the law for much of the 1900s allowed, within certain limitations, prospectors to peg and operate their claims on any farmland. The resulting strife between farmers and miners was inevitable. The noise, the spoiling of the farmland, the spreading of the dumps of broken waste rock and treated sands, and the influx of labourers were unavoidable. Many prospectors were guilty of practices inimical to farming. When the dried grass and bush hindered prospecting, many farms were torched. On many an occasion a lighted candle would be planted down an antbear hole, with a wick at the base of the candle leading to a bundle of dried grass; the prospector would then noisily and ostentatiously leave the area and drive to town. When confronted by the infuriated farmer whose grazing had been burned out hours later, the prospector might reply, with injured innocence, 'Who? Me? But you saw that I spent the day in town!'

Extract from my [the author's] diary: Gwanda District, April 1955

> On the sloping ground of the low hill there is a maze of shallow quarries seemingly following no systematic plan. They remind me of the aimless wandering of caves dissolved in limestone by acid waters that have eaten away at the strata for countless millennia. But this is hard, fine-grained greenstone – ancient lava engulfed several billion years ago by great intrusions of the granite that forms the old foundations of Southern Rhodesia.

The quarries are the work of 'The Ancients', the mysterious people who had followed the richer threads of gold-bearing quartz that streaked the old lavas. As they had pounded the rock faces with their stone hammers, the quarries had grown wider and deeper. When the golden threads had petered out, they had simply moved to another spot. Assays of remnants of these threads left behind ran up to an ounce of gold to the ton.

Thirty or forty years ago, smallworkers had moved in and sunk a gently inclined shaft beneath the old quarries to intersect the gold-bearing ore at depth. The hard, fresh ore was blasted out with explosives, hoisted to surface, and crushed to sand in the five-stamp mill that still rattles away on a low hill close by. The dump of blue-grey sand, discarded after treatment to extract its gold, bears testimony to the thousands of tons that have been treated.

Twelve metres underground, the miners of the new age had blasted their way into a dungeon-like cavity that the Ancients had opened up in their pursuit of the ore. As I gazed into this hole, shining my miner's lamp into its distant recesses, I had visions of near-naked bodies drenched in sweat, pounding the rock faces with their crude hammers and eventually dislodging the slabs that spalled off. The light they worked by would have been a flickering glow from small fires or torches of resinous roots. The air would have been foul, filled with smoke and stinking of garlic from the arsenical vapours given off by the rocks, and of urine and probably faeces. It would have been a foretaste of Hell. From time to time, others would have carefully gathered up the chippings and swept up the dust that collected at the feet of those who worked at the rock face. Placing this in skin or closely woven baskets suspended by thongs from their shoulders or foreheads, they would have used precarious footholds to climb out of the hole and deliver it to those whose task it was to extract the gold particles.

As I stood there, the questions crowded in: Who were these people? Did they of their own accord learn the value of gold in trade with the Arabs who came to the far-away

> east coast in their dhows, or were they forced, as slaves, to fulfil a monthly quota dictated by the local chieftain or regional ruler? Did they sing and joke amongst themselves as they worked, or did their heads droop in abject submission and near-exhaustion?

Extract from my diary: Gatooma District, February 1956

Similar visions of ancient times flitted through my head in February 1956 when I was working in the Gatooma District:

> The mine I am to spend the next few weeks mapping and sampling is a gaping hole 180 metres long and 50 metres wide at surface. It had been abandoned 20 years ago when the near-surface ore was exhausted. Access is now by an inclined shaft that starts to one side of the great pit, and ends on the floor 60 feet below. The rail tracks had been lifted years ago and taken away, so now you slither down on your buttocks, starting small avalanches of loose rock ahead of you. On reaching the bottom, a pair of nesting rock pigeons takes off with an angry clatter of wings. Standing in pools of dirty, greenish water are great boulders that have fallen from the walls of the pit, and now rear up like icebergs in some Antarctic sea. But down here there is no ripple of breeze, and it is fiendishly hot from the sun beating on the rocks.
>
> Remnants of the ore still lie scattered around. It is yellow-brown, thoroughly oxidised, rotted and crumbly – ideal for the practice of direct cyanidation that had once been used to extract its meagre gold content. Using a hand lens, the minute spicules of arsenopyrite can still be seen in it, now weathered to crumbly brown iron oxides.
>
> Where the excavation ends abruptly against a vertical wall blasted out of the rock at its northern end, a gash is exposed, starting at surface and ending halfway down to the floor. It is no more than a slit 45 centimetres wide, now choked with soil and rubble – the debris of the years. Clearly, a slit once filled by a quartz vein mined out by the Ancients, God knows how many centuries ago. It was the

rich gold of a vein that they had followed, ignoring the great tonnage of low-grade ore in which it was embedded, and from which they had had neither the knowledge nor technology to extract its riches.

I ask myself: Who were these people, and who or what drove them to pursue the yellow metal down a hideously dangerous and impossibly cramped slit in the ground like this? Did they squeeze their bodies between the confining walls, or did they use child labour, or perhaps slim-bodied Bushmen pressed into service as slaves?

THE MYSTERY OF THE PREHISTORIC MINES AND RUINS

In his presidential address to the South African Association for the Advancement of Science, Williams (1906) recalled the conclusions reached during the 1891–92 expedition to Great Zimbabwe led by Theodore Bent, sponsored by the Royal Geographical Society. In the ancient ruins Bent believed he saw '. . . evidence of a cult known to Arabia and Phoenicia alike, temples built on mathematical principles . . . and evidence of a vast population devoted to the mining of gold'. An interpretation that followed was that 'Ophir' was not a primary source of gold, but a port on the south coast of Arabia to which the gold was brought and then traded (in other words, an entrepôt centre).

Some, comparing the gold work and carved stone emblems recovered from the ancient ruins, with those found in Egypt and Arabia, had hinted at an age as great as 1 000 to 2000 BC (White, 1903a). In similar vein, Hall (1903) speculated that 'the main walls [of the passages in the ruins], portions of which are believed to have stood some 3 500 years, are, if given the same conditions, quite likely to be standing at the end of another millennium . . .'

F.P. Mennell (1904), prominent in Rhodesian museum and geological circles, was another who was quite unable to accept that the ancient ruins in the territory had been built by indigenous peoples. His argument was based on a comparison of the simple mud huts, in which the local peoples dwelt, with the massive ancient structures of dressed stone laid in intriguing patterns. The absence of bronze articles amongst the artefacts recovered he saw as further support for dating the ruins as earlier than 1500 BC. Furthermore, he noted that El-Masoudi, the Arab historian, depicted the indigenous peoples following the same customs in the year 930 AD that they did today, and, while Arabic inscriptions proved that Arab explorers had penetrated to the interior, there was no reference to any advanced civilisation surviving there at that time. Mennell, however, had no doubts in his mind that gold mining had had a great deal to do with the building of the stone-walled settlements.

A few years later, Revd S.S. Dornan (1915) was to refute these views on the basis of verbal narratives passed on by some very elderly patriarchs, arguing the case for a relatively late age for the stone-built cities, and the abandonment of Great Zimbabwe only in the seventeenth century. His conclusion, bolstered by observations on architectural styles and building methods, was that 'the foreign influence theory must be given up'.

The German explorer, Carl Peters, formerly a highly placed official in the administration of Tanganyika (Tanzania), was probably as influential a figure as any other in linking the kingdom of Monomotapa with the biblical Ophir. His *Eldorado of the Ancients* (1902) describes a two-year expedition from the mouth of the Zambezi River to Tete in the interior, and then southwards to Manicaland via the Inyanga area to Umtali (present-day Mutare). His examination of ancient ruins south of the Zambezi, his view that the physical appearance of some of the indigenous peoples was 'absolutely Jewish', and his philological interpretation of place names convinced him of an ancient Semitic influence pervading the entire area. To him 'Africa' was a corruption of 'Ophir', as were 'Sofala' or 'Sofara'.

The discovery of a pottery statuette recovered 500 kilometres upstream on the Zambezi River, and its subsequent identification by the famous Egyptologist, Flinders Petrie, as a figure of the Egyptian ruler, Tahutmes III (around 1450 BC) was in Peters's view final proof of early exploration of the interior by Egyptians as well. Peters gave an interesting twist to prevailing ideas in concluding that the subjects of the Queen of Sheba had not themselves mined gold for distributing as largesse to her allies, but that she granted mining concessions to them. Amongst these allies were active miners dispatched by Solomon, from whom he collected every three years the gold they had produced.

Regrettably, the biblical references give no clue as to the location of Ophir. All that can be said, on geological grounds, is that the 7 000 talents of silver included in the donation of 3 000 talents of gold, referred to in I Kings, must have come from a source other than the land of Monomotapa. Zimbabwe is not a silver-producing region of any note, and the small amount of silver in the natural gold alloy from that region amounts generally to 20 per cent or considerably less (Eales, 1960). The silver recovered would have fallen far short of the amount stipulated in I Kings.

It was not only the Portuguese who had sought riches in the kingdom of Monomotapa. Jan van Riebeeck, the first Dutch Commander at the Cape, reputedly made several attempts to locate it. Gardner Williams (1906) records that no less than five parties had been sent out on abortive expeditions in the years following 1669, long before the celebrated foray led by Governor Simon van der Stel in 1685. With 56 white companions and a troop of attendants drawn from the local inhabitants, Van der Stel was successful in discovering copper ore in Namaqualand, but the location of Monomotapa remained unresolved. It is recorded that he had plotted on a map his estimate of its location – ironically, close to the present Witwatersrand.

In a report on the tin mines of the Rooiberg area of the Transvaal, Baumann (1919) also speculated that the tradition of mining was introduced by a Semitic people from southern Arabia at some time earlier than 600 AD and that, when they withdrew again from southern Africa, they left behind knowledge that was handed down from father to son over the generations. By the beginning of the nineteenth century, these skills appeared to have been lost, and memories of who the ancient miners might have been had faded. An intriguing reference is, however, made by Rosenthal (1950) to a belief amongst the native peoples along the Limpopo River that in ancient times the area had been visited by men wearing a white cloth around their heads, and travelling on camels.

A wealth of literature exists on this topic, notably from Bent, Thomas Baines, Theal and others, but enough has been presented here to illustrate the beliefs that were fashionable at the beginning of the twentieth century. In the years that followed, however, archaeologists constructed a very different scenario.

TWENTIETH-CENTURY ARCHAEOLOGICAL EVIDENCE

A concise review by Miller (1995) reveals a tradition of mining in southern Africa extending back to the beginning of the first millennium AD. Slags from the smelting of iron ore, and metal samples, show that mining was in progress in Mozambique early in the second century AD, in South Africa in the fourth century, and in Botswana in the sixth. Radiocarbon dating gave a date of 770 AD for charcoal recovered from ancient copper mines at Phalaborwa while charcoal from copper ingots from the tin-mining district of Rooiberg has yielded somewhat younger dates spanning the eleventh to fourteenth centuries.

Indications of early gold-mining activity are by no means confined to the region north of the Limpopo River. Tom Huffman (1996) has outlined a 'Zimbabwe Culture Area' stretching from well south of the Limpopo River to north of the Zambezi, embracing parts of today's Botswana to the west and Mozambique to the east. Within this region, three distinct cultural periods are recognised by him: Mapungubwe (1220–1290 AD), Zimbabwe (1290–1450) and Khami (1450–1820). Mapungubwe is held to have been the largest settlement in southern Africa at the time. Nearby are the ancient remains of K2, and of Schroda, the earliest known Iron Age settlement in southern Africa to yield numbers of glass beads, indicating that it was one of the first to be integrated into the Indian Ocean commercial network. Huffman regards Great Zimbabwe as the economic, cultural and political successor to Mapungubwe.

Graves excavated at Mapungubwe, close to the confluence of the Limpopo and Shashi Rivers, are said to have yielded the remains of a woman wearing more than 100 gold bangles, and some 12 000 gold beads; another grave yielded a bowl, a sceptre and a figure of a rhinoceros made of gold sheet tacked onto a wooden core. Passing reference is also made by Rosenthal (1950: 122) to ancient mine workings in the Pilgrim's Rest area in what was formerly the eastern Transvaal, and to gold-smelting furnaces, slag heaps, and primitive, wooden mining tools described in the 1930s by D.S. van der Merwe, the mining commissioner at Pietersburg in the northern Transvaal (Rosenthal, 1950: 138).

Archaeological investigations sponsored by Gold Fields of South Africa (Spamer, 1996; Allen, 1997) at Thulamela in the Kruger National Park have established that a large stone-built settlement flourished here too, between the fourteenth and seventeenth centuries, with a population of some 2 000 people. Evidence of gold smelting was found, while beads and porcelain collected there imply trade links with China and India.

Pikirayi (2001: 5–36) of the University of Zimbabwe gives a synthesis which is dismissive of any thesis other than one that Great Zimbabwe was built by indigenous peoples. He recalls the research by Gertrude Caton-Thompson in 1929–31, which yielded evidence for occupation of the site as early as the eighth to tenth centuries, but revealed nothing to suggest occupation by non-indigenous peoples. The ruins were interpreted as medieval in age. This supported, in a general way, earlier conclusions of MacIver in 1906 and Dornan in 1915, but no evidence was found to indicate that the growth of the stone city had been founded on the gold trade. Subsequent work by later archaeologists, notably Roger Summers and K.R. Robinson, and the use of radiocarbon dating, has suggested that the technique of building in stone had evolved amongst the indigenous peoples early in the second millennium, but that Great Zimbabwe had fallen into decline before 1450.

The great debate is not closed. Pikirayi (2001: 5–35) scorns interpretations harking back to Phoenician and Semitic influences that still appear in recent literature, but the vexing questions cannot be considered to have been answered in full. What provided the initial stimulus for gold mining, and from whom was the value of gold in commerce first learned by the indigenous peoples? From whom, if there were indeed ancient tutors, had the skills of mining been acquired? The answer to the first question would seem to be obvious. The stimulus must certainly have been the acquisition of a commodity with which to trade for the cloths and trinkets from across the ocean, for as Randles (1975: 85) puts it, 'Contrary to a fairly widespread idea, gold in itself hardly interested the Black population.' Other writers have expressed the same view. The remaining questions are part of a skein as tangled as that woven around Ophir itself.

EXTRACTING THE GOLD FROM THE ROCKS: THE TECHNOLOGY OF THE 'ANCIENTS'

Despite primitive technology, some of the ancient mines within the Zimbabwe Culture Area reached significant depths. The Geelong Mine in the Gwanda district of Zimbabwe is reported (Summers, 1969: 118) to have reached 36 metres below the surface, while an early report by Trevor and Mellor (1908) described ancient workings for copper extending for some 30 kilometres in the Messina district. These had reached depths of 24 metres before being abandoned when the water table was reached. At this depth the ancient miners could no longer have coped with the inflow of underground water.

In the absence of any written records, we can only reconstruct from physical evidence the methods used underground by the 'Ancient

People'. Without explosives, breaking of the ore underground would have been back-breaking labour, using stone hammers and primitive iron tools. It is certain, from numerous findings of firewood piled against old underground rock faces re-exposed by more modern mining, and from traditional oral narratives recorded by Dornan (1915), that the technique of firesetting was known and used. According to Dornan's interpretation, this was introduced by Arab miners.

Although the following description of firesetting by Agricola (1556; translated by Hoover & Hoover, 1912: 118) refers to the sixteenth century in Germany, it does help to conjure up a mental picture of what the procedure entailed, and the hazards incurred during its execution:

> As I have said, fire shatters the hardest rocks, but the method of its application is not simple . . . a heap of dried logs is placed against the rock and fired . . . This force does not generally soften a large portion of the vein, but only some of its surface . . . While the heated veins are giving forth a foetid vapour and the shafts or tunnels are emitting fumes, the miners and other workmen do not go down in the mines lest the stench affect their health or actually kill them.

Agricola (1556: 215) goes on to say:

> There is another illness . . . which soon brings death to men who work in those shafts or tunnels in which the hard rock is broken by fire. Here the air is infected with poison . . . from the minerals, which is driven out by the fire . . . which clings to the upper part of the walls [or] falls down into the pools and floats on their surface . . . Prudent and skilled miners burn the piles of wood on Friday, towards evening, and they do not descend into the shafts . . . again before Monday.

This illness the translators attribute to arsenical poisoning resulting from the roasting of arsenical minerals. *Arsenopyrite*, with the composition FeAsS, occurs widely in Zimbabwean gold deposits, and the same hazard must have taken its toll on these unfortunate workers. On resumption of their work '. . .[the miners] thrust a smaller crowbar into the cracks and so break it down.' Strangely, Agricola makes no reference to the application of water to the heated rock to promote cracking. This was certainly practised in Zimbabwe, as evidenced by broken clay vessels and troughs for water found in old workings.

A brief account of a primitive rock crusher in the vicinity of the Geelong Mine, where a great boulder was rocked on a granite pavement to pulverise the gold ore, was provided by Eales (2001: 69) (see Plate 21). Surprisingly, a somewhat similar primitive device was in use in 1871 at the Eersteling Mine in the northern Transvaal, as shown in a watercolour painting by the artist, Thomas Baines (see Antrobus, 1986).

Figure 49: Ancient trench-like working at Tebekwe Mine, Shurugwe, Zimbabwe, revealed by excavation carried out by mine owner Bernard Beeks in 1989. Close up, tool marks left by the early miners are clearly visible in the side walls of the trench. (Photograph by courtesy of Lorraine Swan, until recently archaeologist at the National Museum, Harare, and Timothy Broderick, formerly with the Geological Survey, Harare.)

Wherever mining was carried out, there will generally be found, close by, places where the ore was crushed in 'dolly holes'. Shallow holes in natural rock pavements acted as mortars in which the ores were pounded with rock pestles, and the surrounding areas are often seen to be worn smooth by the shuffling of feet. The pulverised ore would then have been panned in some sort of vessel to wash out the lighter minerals, leaving the gold 'tail'. For this reason the dolly holes are usually sited near a stream or pond.

Although trivial by today's standards, the total amount of gold that had been mined in what is now Zimbabwe was far from insignificant. The estimate of Summers (1969: 188) was that some 25 million ounces in total had been won, with the most productive period having been 900–1200 AD, followed by a gradual tailing-off through the following centuries. It might be added here that, after years of digging, some mines in the territory would have reached the water table, and for this the early miners would have had no answer. This, conceivably, might have been a factor in the progressive decline in production indicated by the estimates, discussed earlier, compiled by Randles (1975).

Figure 50: Dolly holes in which ores were crushed by the early miners ('the Ancients') at the Formadunga Site, Chiredzi, Zimbabwe. The holes here are located in gneiss, and the local geology suggests that in this instance they were used for the crushing of magnetite-bearing quartzites as a source of iron. Mapungubwe-type pottery is reported to have been found here. (Photograph and details supplied by courtesy of Lorraine Swan and Timothy Broderick.)

Figure 51. Extraordinarily abundant and closely spaced dolly holes at the Macardon Claims, West Nicholson, in the Gwanda District of Zimbabwe. Their close crowding renders it unlikely that all could have been simultaneously in use – there would not have been standing room. Does their arrangement in rows suggest that this was the introduction in prehistoric times of the 'conveyor-belt assembly-line' approach to speed up production? (Photograph taken by Neville Jones in 1939, and passed on by Timothy Broderick of Harare.)

It may be speculated that final cessation of any significant gold-mining activities was brought about by the *mfecane* in around 1840 – the sacking of the country by Ndebele invaders under Mzilikazi. Mennell (1904) and Dornan (1915) do refer, however, to the renowned hunter, Selous, recording that some gold mining was still being carried out by indigenous peoples in 'Rhodesia' and what is present-day Botswana, early in the nineteenth century.

Postscript from [the author's] diary: Bindura, February 1957

> Pushing my way through a splendid grove of trees (mountain acacia?), I came upon a funnel-shaped hole some 20 metres across. Where the hole tapered to the bottom, the remains of

the square shaft still framed in solid timbers could be seen. No Ancient working this, but an ever-widening collapse of ground down the shaft of an old mine that had been abandoned many years before after, one night, the floor of one entire level some hundreds of feet down had broken away along a parting of talc schist, and plummeted into the level below. The miners had paid too little attention to the properties of talc schist, which has strength only a little better than that of a bar of soap. The entire night shift had been trapped and crushed under tons of rock. Exhumation of their remains was impossible, the mine was a shambles, and it was simply closed and abandoned. The headgear and all the plant and machinery had been stripped down and taken elsewhere. Many years' growth of bush and trees now hid the evidence of the old disaster.

It was an eerie experience to be thrusting your way through the bush and suddenly coming upon the most exquisitely dressed 2.5 metre slab of granite, chiselled to shape decades ago by Cornish workmen for anchoring an engine or piece of machinery of some sort. On one side was carved a deep scallop to accommodate a flywheel, and there were holes neatly drilled through the block to receive long bolts for anchoring the engine firmly in place. Mental images of a bustling workforce, and the cheerful banter of the labourers, and of some overseer taking his ease upon the ground while smoking his pipe, contrasted with the dead silence and tangle of bushes that now grew there.

Moving away, I had stopped briefly in a small, grassy glade to write comments in my notebook, when I felt my feet breaking through a superficial crust of soil and grass and sinking into a pit of sticky, black tar. This had clearly been a pit into which the tar produced as a by-product of the suction gas engine, running on green firewood, had been discharged. With visions of being trapped and sinking like some prehistoric animal in the tar pits of Venezuela, and my fossilised remains being dug out thousands of years later, I wrenched my feet clear and stumbled away to safety . . .

35
Mineral Deposits: Discoverers' Elusive Rewards

As he often did in other topics that he wrote about, Agricola taunted his readers about the ethics of chasing gold, quoting from older works:

> 'Gold and silver are injurious to mortals; gold is the source of crime, the plague of life, and the ruin of all things . . .' (*Phocylides*);

while offering an opposing viewpoint closer to his own:

> 'The greatest rewards come from gold; by gold love is won . . . by gold is justice bought; the law follows the track of gold . . .' (*Propertius*)
> *Agricola (1556: 9–10)*

> '. . .the born prospector rarely abandons his occupation, although so few ever attain a material reward in any way adequate to the energy they have put into its pursuit.'
> *E.T. Mellor (1922), presidential address to the South African Association for the Advancement of Science*

Whichever philosophy offered by Agricola accords with your own, it is probably true to say that most people assume that immense wealth accrues more or less automatically to the discoverers of mineral deposits, or to the owners of the properties on which the deposits are found. History shows that this has not generally been the situation in the past.

As a famous case in point, the Main Reef Leader of the Witwatersrand goldfields was first discovered on the farm Langlaagte, where Johannesburg now stands, on a date taken to be 7 February 1886 (Werdmüller, 1986) by prospectors George Harrison and George Walker. (Because of the disposition of the magisterial districts at that time, the goldfields were first known as the Heidelberg Goldfields.) Harrison received a mere £10 for his discoverer's claim, and his partner, George Walker, £350 for his. Harrison then disappeared from the scene, and his subsequent history remains a mystery, although there is a legend that he went to Barberton where he died of fever somewhere in the hills (Bulpin, 1955: 67). The record of later deals involving Harrison's solitary claim has been sketched by Fivaz (1988). The buyer, F.W. Marsden, sold it for £50 to Alfred Hepple, who passed it on to the Little Treasure Gold Mining Co. for £1 500 and shares to the value of £150. The claim was in turn acquired by the Northey Gold Mining and Exploration Company for £2 000 in cash and £5 000 in shares. The original owner of the farm, the widow Petronella Oosthuizen, received £12 000 when she sold out to J.B. Robinson (Gray, 1937: 89). The original discoverers therefore received virtually nothing.

Rapid development and tracing of the reefs followed the discovery by Harrison and Walker, with phenomenal grades being recovered at surface and down to shallow depths in the earlier days. Gray (1937: 191) records that the Wemmer Company recovered 295 ounces of gold from 35 tons of ore (an average of 8.4 ounces per ton), and that more than 500 trial crushings of ore yielded an average of 3.25 ounces per ton. With mercury amalgamation, the recovery was probably in the order of only 60 per cent or less, although Fivaz (1988) suggests a figure of 75 to 80 per cent. The original average value of these ores tested was thus probably close to 5 ounces per ton – enough to make the heart of any prospector, then or now, race with excitement.[7]

Other tales of similar genre are recounted by the historian, Eric Rosenthal (1950). His father, a dealer in shares and mining ventures, told of the farm, Rietfontein, changing hands in 1883 for a saddle and a bridle, but being the centrepiece of a £1.6 million deal in 1934.

Incomplete records show that during the period 1887–1984, 1 080 million tons of ore from the Central Rand goldfield were crushed for a recovery of a little under 9 million kilograms (290 million troy ounces) of gold, averaging 8.3 grams per ton (Werdmüller, 2001). The richer near-surface ores had thus carried ~20 to 30 times more gold, per ton of ore, than the average ore mined in the following 90 years.

On the farm, Vooruitzigt, owned by the brothers De Beer in 1871, stood Colesberg Kopje, the site of the future Kimberley Mine. E.H.L. Schwarz (1908: 59) records how the younger brother sold his half of the farm for £1 000, and the elder brother his for £5 600, but appeared not to have been over the moon at the sale, commenting that if he should ever buy a farm again he would ensure that there were no diamonds there. In 1891 the sale of diamonds recovered from the De Beers and Kimberley Mines fell just a little short of £4 million. As an early pioneer, Henry Kisch, put it: 'There is a curse on all of us early comers. I was on the diamond diggings before Cecil Rhodes was ever thought of . . . none of us who was there has a penny . . . if you come too early, you get nothing' (Rosenthal, 1950: 17). No worm for the early bird here!

Not all landowners came out on the losing side. Rosenthal recounts how Willem Prinsloo, a farmer whose only wish was to be left alone and in peace, reluctantly sold Modderfontein in 1886 for £70 000 in shares, £30 000 in cash, and a prize cow, to escape from the importunate prospectors. He then moved to Kaalfontein, merely to suffer further unwelcome pestering there. This persuaded him to sell out, this time for £100 000, and move to yet another farm 65 kilometres from Pretoria. Incredibly, the peace and solitude he craved still eluded him, and he was importuned into selling half of this property – this time the site of the future Premier Diamond Mine – to Thomas Cullinan for £80 000. In another version of this tale, Kanfer (1993: 159) states that it was Prinsloo's children who sold out for £52 000. Whichever version is correct, immense wealth flowed from the transaction. Within five years, diamond production had reached 1.89 million carats per annum, valued at £1.8 million, and by 1930, 28.46 million carats had been recovered, fetching £32.39 million.

36
WITWATERSRAND GOLDFIELDS: EARLY SCEPTICISM, PREDICTIONS AND CRISES

'When I was on my journey home in the steamer from the Cape, a passenger from Natal . . . said that he considered it perfectly fair and legitimate to represent that pieces of gold quartz from Australia or elsewhere, were found in the South African gold-fields.'

James Gregory (1868b)

'No doubt we shall hear more about these alleged discoveries.'

Die Volkstem *(28 June 1886), an influential newspaper of the former Transvaal*

'I do not consider the thing is a "boom" . . .'

Cape Times *correspondent (12 July 1886)*

One overseas geologist gave it as his opinion, four months after discovery, that the deposits would not continue for more than fifteen feet below the surface.

R. Borchers (1964: 8)

The many abortive attempts to establish profitable gold mines in the Transvaal during 1850–86 had led, quite understandably, to scepticism about the initial reports of gold at Langlaagte in 1886. Several factors were in play here. The first was rooted in the sparse geological knowledge at that time – failure to appreciate the potential importance of ancient conglomerates as carriers of gold. The general belief was that gold is associated with quartz reefs – despite the success of the rich Tarkwa goldfield (Gold Coast of West Africa) in 1881, where deposits very similar to those of the Witwatersrand occur.

A second factor was the attitude of the government, which was decidedly averse to an influx of *uitlanders* ('foreigners') that might follow any significant discoveries, and who might vie for political control. Werdmüller (1986: 11) quotes from the research of James Gray that John Henry Davis reported in 1851 to President Pretorius that he had discovered gold on the farm on which Krugersdorp now stands – and was promptly expelled from the country. In 1853, P.J. Marais was allowed to prospect for gold but, under pain of death, was warned to divulge any discoveries to no one but the government.

The subsequent discovery and development of the Witwatersrand deposits in 1886 was to reveal, despite the scepticism with which the first reports were received, a treasure house unmatched by anything similar in the world. Within four years, however, there emerged a technological problem which left the entire gold-mining industry foundering. The future of the City of Gold had hung in the balance.

THE OXIDATION CRISIS OF 1890

'Doom and despondency seized the town of Johannesburg [in 1890] where just about all the inhabitants were dependent on the prosperity of the mines. Gold shares were sold on an unprecedented scale, companies crashed, and mines closed down. Shares could not be sold for 5 per cent of their previous value.'

C.E. Fivaz (1988), presidential address to the South African Institute of Mining and Metallurgy

'There was clearly no future for mines that were not drawing ore from the rich Main Reef Leader.'

A.P. Cartwright, as quoted by Fivaz

'Grass will grow in the streets of Johannesburg within a year.'
Sir Percy Fitzpatrick, as quoted by Fivaz

The crisis that prompted these gloomy predictions was brought about by a commonplace, but purely geological phenomenon – the change from rocks that were highly oxidised near the surface, where exposed to the effects of air and water, to unaltered, sulphide mineral-bearing ores as the depth of mining increased. In thoroughly oxidised surface ore, the gold particles are amenable to simple amalgamation with mercury, which was the recovery process employed during the first three to four years of operations. Secondary enrichment (solution, redistribution and redeposition of the gold) is also encountered in near-surface ores, to which the occurrence of gold nuggets, occasional well-formed crystals, and phenomenally high grades in the earliest days may be attributed.

Figure 52. An old stamp mill used for crushing gold ore on the early Witwatersrand mines. The vertical rods carrying the heavy stamps were lifted by engine-driven rotating cams and allowed to drop onto the ore which was introduced into the mortar box at the base of the structure. A constant stream of water was used to flush out the crushed material.

In these first years the rock was crushed by stamp milling. Stamps weighing one-half to two-thirds of a ton were dropped some 100 times per minute by a rotating cam onto the ore fed in. Several hundred stamps might be employed at a mine, with each stamp crushing five tons per day.

Mercury was added to the mortar box, and the slurry of water and crushed rock issuing from the batteries then led over copper plates coated with mercury to trap the free gold particles and gold-mercury amalgam. A concentrate of heavier minerals was also collected on various types of shaking tables, or corduroy mats. In an aside, Franklin White (1903a) draws an analogy between this technique and the use of a sheepskin fleece in ancient times: '. . . the Grecian fable of Jason and the Golden Fleece is probably a poetical reference to a successful raid on a community of alluvial gold washers'.

In unoxidised ores encountered at depth, the gold particles are generally in contact with or encased by sulphide minerals, chiefly of iron (pyrite, or FeS_2 (see Plate 22)). Amalgamation is appreciably less efficient in such ores, and this dropped the percentage of gold extracted from approximately 70 per cent to less than 50 per cent. This was the crisis that faced the industry (Fivaz, 1988) and cast an ominous shadow over it.

It was pure serendipity that, in 1887–88, the MacArthur-Forrest process was developed by J.S. MacArthur, a Glasgow-trained chemist, and the Forrest brothers, both medical doctors. They showed that very dilute solutions of potassium or sodium cyanide, in the presence of air, dissolve gold and silver in preference to base metals. The gold and silver taken up in solution could then be precipitated by the addition of zinc shavings.

A pilot plant for treating ore by the new process was erected on the goldfields in May 1890, and a large group of metallurgists and engineers attended a demonstration. The results were electrifying. It was established that a combination of amalgamation and cyanidation could achieve 90 per cent extraction efficiency. In the words of W.A. Caldecott (1903), '. . . . the incredulity with which the results claimed were first received . . . was soon changed to a general acceptance of the merits of the new process'. The words of Torrente (1904) were even more exuberant, stating that the cyanidation process 'converted many gold mining companies . . . on the verge of bankruptcy into most flourishing dividend-paying concerns'.

The result was a complete change of outlook on the future of the industry. It was now possible to re-treat the still-rich tailings accumulated over the previous five years, and the prospect of deep-level mining became obvious. Fivaz notes that within twelve months all ground south of the reef outcrops was pegged 'to a depth of three miles', and that 'the complete turnabout in the fortunes of the gold-mining industry could be attributed almost entirely to the MacArthur-Forrest cyanidation process'. Within the ten-year period 1889–98, gold production increased just a little less than elevenfold.

Amongst other historical documents collected by Professor J. Moore at Rhodes University is a letter handwritten in January 1894 by an engineer, Tom W. Wood. This tells of various gold-extraction processes that were tried as alternatives to cyanidation: treatment with chlorine, bromine, or aqua regia after the roasting of concentrates in furnaces. All were found to be too inefficient, too expensive, or simply too dangerous. The letter comments that the use of bromine, in particular, was condemned after the killing of a number of workers.

The change in fortunes following the introduction of the MacArthur-Forrest process emerges from the 'before-and-after' comparisons drawn by Caldecott (1903). He records that, during 1889, 'piles of sands and slimes around the battery were regarded simply as a source of annoyance . . . so little was the latent wealth they contained realized that anyone so desirous of removing them was welcome to do so'. Typical tailings that had been discarded as waste were 'about 10 per cent of 12 dwt/ton concentrates, 65 per cent of 4 dwt sands, and 25 per cent of 2.5 dwt slimes'. In subsequent years it became a rule of thumb that ' . . . a Company relies upon its battery returns for its working costs and upon its cyanide plant for its profits'.

An aside offered by Torrente (1904) suggests that Providence had been kind in the matter of the sands, which had accumulated on site, being still available for re-treatment in later years because of the flat nature of the terrain: 'If a good fall of ground had existed, or some friendly creek or river had been near by, these tailings would have disappeared, and an enormous amount of gold therefore lost.'

It is a supreme irony, not to say gross injustice, that the inventors of the MacArthur-Forrest process gained relatively little financial reward for their efforts. In 1896 the validity of the patents was contested in the High Court by the Chamber of Mines, and the patents and royalties payable were cancelled on technical grounds.

A comment worth repeating here is that of P.J.D. Lloyd, the then Director of the Metallurgical Laboratory at the Chamber of Mines, who in 1978 wrote: '. . . the process has evolved over the past 88 years to its present level of high efficiency . . . it may well be an inefficient plant that reports a recovery below 99 per cent a decade from now.' As mines deepened and approached the 4-kilometre mark, levels of gold recovery such as this would be demanded to offset the resulting increase in operating costs.

We might also ponder whether the following prediction could have been made without the introduction of more efficient gold extraction processes: 'The known reserves would last for another 80 years at the current annual production rate' (Frimmel & Minter, 2002). Without the technology to maximise gold recovery, great reserves of ores of marginal value could not have been considered economically viable.

37 WITWATERSRAND GOLD: WHERE DID IT COME FROM?

'. . . all geologists who have visited any of the mines of that area [Far East Rand] will admit that the correspondence there everywhere apparent between the occurrence of the gold and purely sedimentary features in the conglomerate, in which it is found, is truly remarkable.'

P.A. Wagner (1917), University of Pretoria and South African Geological Survey

'A large proportion of the gold in Witwatersrand deposits has retained its detrital features . . .' and 'there is no convincing evidence that large amounts of gold or uranium were [introduced] in solution.'

D.K. Hallbauer (1986), Chamber of Mines Research Organisation

'Significant correlation between gold and detrital zircon distribution, as well as sedimentary [features], preservation of detrital micronuggets in places, and absolute age data provide evidence for a detrital origin of the gold, which was derived from the weathered products of 3.1 to 2.9 Ga [billion year] greenstone belts.'

H.E. Frimmel and W.E.L. Minter (2002)

'. . . it may be said that the theory of subsequent infiltration [of gold into the conglomerates] is preferable to the others . . .'

F.H. Hatch and G.S. Corstorphine (1905: 146)

'I must confess right from the time I first formed any opinion as to the origin of the Rand gold, I could accept only a hypothesis of hydrothermal introduction and could find no compulsion in the hypothesis of detrital origin'; 'these conglomerates . . . have actually served as the channelways for the entrance, and the receptacles for the deposition, of vast quantities of unquestionably introduced mineral material . . .'

L.C. Graton (1930: 4), Harvard University

'It is abundantly clear from first principles that the uraninite, and therefore the gold, is not of placer [detrital] deposition.'
C.F. Davidson (1965), University of St Andrews

'. . . we present the results of an extensive geological study of Witwatersrand rocks which show that the gold (and associated uranium) mineralization is hydrothermal in origin . . . We find that the gold precipitated as a consequence of interactions of the fluid with shale-derived hydrocarbons present within the basin.'
A.C. Barnicoat et al. (1997)

Amongst the many controversies that have simmered in South African Earth science, that over the origin of the gold in the Witwatersrand conglomerates (the 'bankets', see Plate 22) has been as heated as any. The simple question here is: where did this enormous amount of gold in the Witwatersrand Basin come from? The question cannot be brushed aside as something of academic interest only – it is important because to it are linked the strategies that will be necessary to keep the mines in production for as long as possible into the future. After 120 years of exploitation, the unmined gold that remains is considered still to represent some 40 per cent of the world's available store of the metal.

Even after twelve decades, irreconcilably opposed views about the original source of the gold are still held. In the random sample of opinions presented above, the first three extracts stress the evidence that the gold was washed into the sedimentary basin as fine-grained particles of metal. The opposing view, implicit in the final four quotations, is that the gold was introduced by mineralising solutions that permeated the sedimentary succession.

THE PLACER (DETRITAL) THEORY OF ORIGIN

One school of thought argues that the gold was released by the erosion of ancient rocks ('schist belts' or 'greenstone belts') of the type in which gold occurs as quartz veins, lodes and impregnations in the Barberton and Murchison Range areas, and in Zimbabwe (see Plate 23). It is held that this gold was swept into the great Witwatersrand Basin by rivers draining from the north and west into a generally shallow sea. The action of waves, wind and even of moving glacial ice has been recognised in these very old rock strata.

This is the *Placer Hypothesis*, which assigns a detrital or mechanical origin to the gold in the deposits. The characteristic association of small gold particles with larger pebbles in the conglomerates is held to be a natural consequence of the very high density (specific gravity) of gold – a metal nineteen times heavier than water, and nearly twice as heavy as lead.

THE ALTERNATIVE: THE HYDROTHERMAL THEORY

The opposing point of view is that the gold was introduced into the Witwatersrand rocks by hot gold-bearing solutions (*hydrothermal solutions*) invading them from some outside source. Amongst the most outspoken early advocates of this concept were Professors R.B. Young (1909, 1911, 1913) in Johannesburg and L.C. Graton (1930) at Harvard University, followed, amongst many others, by C.F. Davidson (1965) in Britain and, more recently, by Barnicoat and his co-authors (1997). To account for the mineralisation favouring the conglomerates above all other rock types, it is argued that they were more porous, and thus more permeable, than other rocks such as the ancient shales and the sandstones that were recrystallised to compact quartzites, and that it was this property that had channelled the gold-bearing solutions into them.

It is not out of place to note here that other variations on this theme were put forward shortly after the discovery of the goldfields. In 1888, W.H. Penning had suggested that the gold and abundant pyrite (FeS_2 or iron sulphide) in the ores were chemically precipitated from the same ocean waters that had mechanically laid down the conglomerates. In 1896, Professor L. de Launay had elaborated on this theme. The reasons for rejecting this approach are fairly obvious – such a process of precipitation from a large volume of water would not explain why the conglomerates were the preferred rocks onto which the precipitates settled, while others above and below remained relatively barren.

In his anniversary address as President of the Geological Society of South Africa in 1911, Professor R.B. Young reviewed the pros and cons of the conflicting hypotheses in words that have a distinctly poetic ring. Musing on what might have been the vista seen by some imaginary being 'strolling at low tide over the wide stretches of shingly beach of the period when the Main Reef Series was being formed', he asked:

> [W]ere the pebbly beds all spangled over with fresh glistening particles of pyrite, or were they dark with black iron sands? Did they contain such an abundance of gold particles as would make the modern beach miner weep for joy, or were they practically barren of that metal? If we are to accept the theories of some writers, that beach was not the most salubrious, and the air was already tainted with mephitic exhalations from decaying sea-weeds, while others would have us believe that the waves that lapped the beach held in solution the future wealth of the Rand (Young, 1911: xxii).

In these words he contrasted a simple placer model with others favouring the conversion of black beach sands to auriferous, pyrite-bearing quartzites, or perhaps the precipitation of gold and sulphides from the sea by the release of evil-smelling hydrogen sulphide from decaying organic matter. Foremost in his mind was the question whether the countless tiny spheroids of pyrite that characterise the ores were original water-worn pyrite pellets swept into the basin, or the product of chemical alteration to iron sulphide of the iron-oxide grains of the black sands.[8]

In a remarkable turnaround six years later, Young, the committed infiltrationist in his earlier years, discarded these convictions and crossed the floor to align himself with the placerists. In his book, *The Banket* (Young, 1917), the concluding chapter concedes (probably with reluctance but inimitable honesty) that 'with the data at present available, the writer is inclined to accept the placer theory'.*

THE PROS AND CONS

The introduction of the gold into the Witwatersrand Basin in the first instance by mechanical processes (current action) is favoured today by most geologists actually working in the mines. The

* For those who might be intrigued by Young's change of faith in 1917, it was brought about by several lines of evidence. Firstly, the earlier work of E.T. Mellor, of the South African Geological Survey, had shown that there was a definite correlation between large-pebble size, pebble orientation, robustness of the conglomerate layers, and the gold content of richer areas of the reef. The most gold is found associated with sedimentary structures that would have formed efficient traps for transported heavy particles.

Secondly, the occurrence of gold in well-mineralised pyritic quartzites, where there were no pebbles in the ore, but which held rich concentrations of heavy detrital minerals such as chromite and zircon, persuaded Young that these quartzites were merely the equivalent of the matrix to the gold-bearing conglomerates – conglomerates without pebbles. The claim that the introduction of gold had been controlled by the superior porosity of conglomeratic rocks therefore fell away.

Finally, his earlier belief in the gold-precipitating action of some unidentified and unspecified mineral that he believed would one day be found in the ores had to be abandoned when no evidence for the existence of any such substance could be found, even after years of work.

correlation between mineralisation and specific sedimentary structures and textures of the host rocks is a compelling argument. By contrast, the hydrothermalists tended, in earlier days, to be mainly those who drew conclusions from microscopic and chemical examination of the ores and host rocks. What confuses the issue is the textural evidence, seen under the microscope, that the overwhelming bulk of the gold particles are intimately intergrown with sulphide minerals, and this is best explained by the gold being precipitated from solutions – the present shapes of the gold particles only rarely resemble those of typical transported debris.

Out of this evidence evolved the belief that the great bulk of the originally detrital gold particles were dissolved and then reprecipitated in the same ores with little spatial dispersal, as a result of rising temperatures and pressures during deep burial of the rocks within the Basin. The conclusion is that the textural features of the ores, suggesting a hydrothermal origin, were imposed only at some later stage on gold grains originally trapped in the ores by sedimentary processes. This is the *Modified Placer* or *Reprecipitation Hypothesis*, without doubt the most convincing answer today.

History reminds us that this interpretation, involving dissolution and redeposition, had been mooted as early as 1897 by G.F. Becker in a report to the United States Geological Survey, and again by Professor Prister (1898b) in South Africa, but had been belittled by Hatch and Corstorphine (1905: 143): '. . . there is little in the character of the conglomerates that supports such an explanation . . .'

A flaw in the Hydrothermal Hypothesis is highlighted by the question: Where is the source in which the solutions might have originated? The underlying granites upon which the Witwatersrand rocks rest are disqualified as they are older than the sediments. An answer that had been seized upon by Young in 1909 was conjured up in the final line of his paper: 'I think it is probable that the acid intrusives [quartz porphyries] are the original source of some of the gold in the Main Reef Series.' Others have proposed the Bushveld Complex as the ultimate source, but this is in reality not host to any known gold deposits of any significance. A different approach was developed in 1965 by Davidson, who theorised that the gold and associated uranium were leached out of the overlying lavas and sediments (of the Ventersdorp and Transvaal Supergroups) by groundwaters that had become heated as they penetrated into deeper levels of the crust.

A fresh look at this old problem was offered by England and his co-workers in 2002. A curious feature of the Witwatersrand ores is the presence of myriads of tiny pyrite grains with rounded (spheroidal) shapes. Here the question arises – was this a natural consequence of abrasion of detrital pyrite crystals as they bounced along the river beds or beaches during transport from distant sources, or was this due to a chemical replacement of original rounded particles of other minerals in black (iron oxide-rich) sands and other detritus? Using the techniques of chemical etching and electronic scanning with modern instrumentation, England et al. have shown that the pyrite grains have in reality endured a traumatic history. Original features that record how the crystals grew, and were then fractured and abraded during transport, are still preserved in individual pyrite grains.[9] This evidence strongly favours a detrital origin for the pyrite and, by implication, the grains of gold in the ores.

What would appear to be a nail in the coffin of the Hydrothermal Hypothesis has been provided by radiometric dating of the gold and pyrite grains themselves. From this – the decay of an isotope of the element rhenium (^{187}Re) to osmium (^{187}Os) – Jason Kirk and his co-workers (2002) have assigned an age of 3.03 ± 0.05 billion years to the gold and pyrite particles from the Vaal Reefs Mine. This age is greater than that of the host rocks themselves (2.89 to 2.76 billion years) and, if correct, would prove that the gold was just another component of older sedimentary detritus swept into the Basin.

All this argument about the origin of the gold has greater significance than mere academic interest. As the review by Frimmel and Minter (2002) points out: 'If the modified placer model applies, the past exploration strategies, which are largely based on the distribution of certain sedimentary [layers], will continue to yield the best return. If the hydrothermal model applies, high gold yields should also be expected from structures that cut across [the layering].' Thousands of kilometres of underground exposures in the mines do not suggest that the latter situation exists.

38
THE BUSHVELD COMPLEX: THE RIDDLE OF ITS LAYERING

'... the Bushveld Complex is too big, too puzzling and too little studied to write the definitive description yet. What picture might we construct from all the different studies that have already been made? There are so many ...'

H. Eales, extract from diary of 1985

The conundrum here lies in understanding why the enormous Bushveld intrusion crystallised as a succession of superimposed layers, like a stack of pancakes (known as the Rustenburg Suite) some 7 to 9 kilometres thick, rather than as an undifferentiated, homogeneous mass. These layers change composition either abruptly from one to the next, or by gradation in their mineral constituents. Individual layers may be followed for hundreds of kilometres, and persist to greater depths than have thus far been plumbed by deep exploratory drilling. The question that taunts us is: how did this come about?

In the mid-1980s, when the Council for Scientific and Industrial Research had sponsored the National Programme for Research on Layered Complexes, groups from several different universities, mining houses and other research groups were brought together to address this question. For readers who have never ventured into the deeper underground levels of a platinum mine, the following extract from my diary may offer a useful word-picture of a typical visit to collect sample material for later analysis.

> You step into the cage suspended by its heavy steel cable that loops over the great wheel at the summit of the shaft headgear. You are wearing a heavy sheepskin jacket on top of your overalls, for this is the high plain of the Bushveld a few kilometres to the north of the Magaliesberg Range, and the air is frosty at this early hour in Winter. You are clumsy in heavy boots and thick webbing belt around your waist, with its heavy battery-pack slapping your buttocks as you walk. Clipped to your miner's hard hat is a lamp, with its electric cord snaking over your shoulder. You are one amongst a dozen others in this compartment of the cage, setting about their tasks.
>
> Gates crash shut and bells ring, and the cage begins to drop, slowly at first but rapidly gathering speed until the lights at intermediate stations down the shaft flicker only briefly as you hurtle past them. The air becomes heavier and you swallow hard to relieve the pressure building up in your ears. The cage rocks gently as it slithers down its guide rails. Eventually the downward plunge slows, and your body is heavy as your feet press against the steel

floor of the cage as it decelerates. The cage comes to a halt, yo-yoing gently up and down with the stretch of the steel umbilical cord that links you to surface. Then, again, bells ring and steel-mesh gates clatter, and you step out onto the concrete floor of your station. No claustrophobic tunnel this, but a hall-like chamber blasted out of the solid rock, brightly lit by strip-lights reflecting off the white-painted walls. Stretching away into the distance is the main haulage, wide enough to accommodate a bus, and you have a momentary feeling that you are entering a tube station on the London underground. It is difficult to accept that here, in the middle of southern Africa, you are walking below sea level.

Ahead of you is a half-hour walk or, if you are lucky, a cadged 'bicycle-ride' on a frame with wheels running on the steel rails laid for the locomotives and ore trucks. The fresh air pumped in by giant fans blows hard enough for you to clap your hands to your hat to prevent it from blowing away, and it carries the chill of the frosty air above. Eventually you reach the development end and clamber over broken rock and pools of water to the face where the roar of the jackhammers is an ear-shattering assault that makes conversation impossible. Here the air is stale, heavy and hot. It stinks of the fumes from blasting, tainted with the faint sourness given off by sulphide minerals freshly exposed to the moist air. Your lamp carves a cone of white light through the fog squirting from the jackhammers, but the rock face is what you have come to examine and everything else is irrelevant – even the throb where you barked your shin against a loose rock.

'Is this where research leads you?' you ask yourself, but the glow of satisfaction as you see the rock face is all the answer you need. At your feet is the greyish-white anorthosite that forms the lowermost rock layer of the sequence you have come to map. On it lies the coarse-grained mass of interlocking greenish crystals flecked with the metallic yellow of sulphides and dull white of feldspars that constitute the Merensky Reef – the World's foremost source of platinum. The Reef is separated from its floor of anorthosite by a thin parting of blue-black chromite crystals that twinkle and throw back pinpoint flashes of light as your lamp plays over them. The parting emphasises the curiously dimpled surface that separates the Reef from its floor – reminding you of the dimpled pattern of moulded fruit-trays that are used in fruit markets to store apples. In reality, a pattern produced by re-melting of the solid floor when it was corroded by hot magma streaming over it 2 060 million years ago.

Another thin line of chromite crystals, a couple of centimetres thick, separates the top of the Reef from the overlying layer of grey-green Merensky Pyroxenite, composed largely of crystals of the mineral *bronzite*, a member of the *pyroxene* mineral group. All of this magnificently displayed in three dimensions in the rock face and the side walls of the excavation. You see detail here that no flat, two-dimensional outcrop at surface or an effectively one-dimensional rod of borehole core could reveal.

You measure and carefully map the exposure, photograph it with flash for your records, and then prize your samples from the rock face, labelling each one and dropping it into a sample bag. At this depth, the rocks are hot to the touch. Then the job is done and, covered in sweat, you set off to trudge back to the station, lugging your samples that will later be used in microscopic, chemical and isotope analysis in the spotless environment of the laboratory. You wait for a cage to take you back to surface, rubbing shoulders with the miners finishing their shifts and gently teasing you with, 'Hey, how's the rock doc today?' and taunting you with stories peppered with adjectives that would make a bartender blink. When they realise that bad language is to you water off a duck's back, they grin amiably and offer you a cigarette.

As you wait, the question that goes round and round in your mind for the hundredth time is: 'What caused this exquisite layering?' For what you have seen this morning, and so many times before, is no more than one page in some thick book – only a tiny part of a greater structure that extends downwards for more than a kilometre, and then above you for several thousands of metres more – thousands of layers, some no thicker than a pencil line, others metres or tens of metres thick. This mass of rock being burrowed through is the Bushveld Complex, once a pool of molten magma injected into the crust when it was younger. You have your own ideas about its origin, but have to accept that here is a problem that a hundred years of meticulous research by some of the World's leading scientists has not yet cracked. What concentrated the platinum-group metals into certain layers to create the greatest reserve of this metal on Earth? Or chromite crystals into other layers to amass the World's largest store of mineable chromium ore? Or stored 60 per cent of the World's vanadium resources in yet other layers here?

As you wait, an old Indian folk tale comes to mind. There were once three holy men, blind from birth, who asked the people of their village to lead them to an elephant so that they could form a mental image of the huge beast. The first was led forward, and felt the tail brush his face. Grasping it in his hands, he squeezed it and then called out: 'Behold, the elephant is like a rope!' Then he walked away, content in his knowledge. The second blundered into one of the massive legs. 'No,' he cried, 'the elephant is like a tree!' The third, on grasping a huge ear as it flapped the air, called out, 'Not so – the elephant is like a fan!' The moral of the tale? Each man had his own opinion, but the elephant is too big for one blind man to comprehend. None of the descriptions was actually wrong, merely misleading, and only by combining a great many more 'touchings' would the full picture eventually emerge. As is the case with the Bushveld Complex.

Figures 53 and 54: Layering is not equally spectacular in all parts of the Bushveld Complex. The upper photograph shows well-developed rhythmic layering, with repeated upward gradation from dark (pyroxene-rich) rock to lighter (feldspar-rich) material in the upper part of each layer. The lower photograph shows thin, discontinuous lenses defined by variations in the amount of white feldspar present in the rock face.

THE SEARCH FOR ANSWERS

Layering in igneous rocks represents a natural rock structure that can be peered at from every angle, have every aspect measured in meticulous detail, be dated by radiometric methods, and be chemically analysed using any number of techniques, but will still not surrender a key that tells precisely how it evolved. The riddle here lies in interpreting a feature that formed more than 2 billion years ago, at a depth that we cannot and will probably never reach, and evolved under conditions that can never be duplicated experimentally.

A wealth of literature offers different solutions to the problem, with only a few of the following concepts capable of being ruled out on a basis of sheer improbability. It is not crucial that the technicalities be explored here: my aim is merely to illustrate how diverse the interpretations are.

Main theme: Layering forms under the influence of gravity

Crystals form and then settle through the cooling liquid magma under gravity. In so doing, they follow Stokes's Law, and are sorted into layers according to the size and density of crystals. This was the most widely accepted view in earlier days.

The sinking of crystals through a medium such as molten rock is an interpretation that goes back to Leopold von Buch in 1824, and appears also in the writings of Charles Darwin in 1844.

Main theme: Layering develops only in large, slowly cooled magma chambers
'When . . . there is scarcely a mass of gabbro of notable dimensions without its banded portion there is no escape from the conclusion that banding originates during slow crystallization.'
N.L. Bowen (1928)

Laboratory-scale proof of crystal settling during slow crystallisation is attributable to the work of Norman Bowen (the great American experimental petrologist) in 1915. His *Evolution of Igneous Rocks* (1928) shows that he followed this as a basic creed throughout his life. Generations of geologists have adopted the same philosophy.

Main theme: The layers were originally sedimentary rocks
'Layers are formed by the transformation of layered sedimentary strata into rocks of igneous appearance and composition.'
S. van Biljon (1949)

In mitigation of Van Biljon's (1949) somewhat bizarre ideas, his transformationist approach stemmed from his claim that 'in certain places . . . bedded structures could still be recognised [in the layered rocks]'. Structures identical to those found in typical sediments do indeed occur here, but he had discounted their being produced by analogous processes (active currents and crystal settling) in large pools of molten rock.

Main theme: Each layer represents the intrusion of a different type of magma
'. . . it was found that each of the constituent rock-types of the mafic portion of the Bushveld Igneous Complex, namely pyroxenite . . . anorthosite, norite . . . gabbro, ferrogabbro, dunite and magnetite, represents a separate intrusion.'
F.J. Coertze (1958), South African Geological Survey

Coertze's view (1958) that each layer represents the intrusion of a different liquid batch runs into severe problems of a different sort. It requires precisely the same sequence of layers to have been injected into different limbs of the Complex hundreds of kilometres apart – a highly improbable scenario. As Cousins (1958: 393) asked: 'Can one conceive of a series of intrusives, each one of which remains on its own horizon over distances of hundreds of miles without ever changing this horizon?' Another formidable problem is the implied existence of chromite magmas. To be fair, Coertze himself made no categorical statement about this, but an earlier proponent of similar ideas did – Edward Sampson (1932) of Princeton University, USA. Such a proposition is quite untenable: chromium has very limited solubility in basaltic magmas (Eales, 2000, 2003). Furthermore, the refractory nature of chromite is revealed by the high temperatures (1 500 to 2 000 °C) needed in arc furnaces to achieve even partial melting of it – well above those characteristic of Bushveld-type magmas.

Main theme: The same magma was repeatedly intruded in a pulsatory manner
'. . . each unit reflects a major renewal of the composition of the magma by a large introduction of fresh liquid in place of that which had partially crystallized.'
T.N. Irvine and C.H. Smith (1967), Carnegie Institute, Washington

The elegance of this model of Irvine and Smith (1967) remains as persuasive now as it was when their study of the layered, 270 kilometre-long Muskox intrusion in Canada revealed a rhythmic repetition of mineralogical and chemical features through some 2 000 metres of vertical section. This could best be explained by successive pulses of intrusion of hot liquid, partial crystallisation of each batch, and bodily displacement of the liquid residues by further pulses initiating new cycles. The model is certainly applicable to parts of the Bushveld Complex (Eales & Cawthorn, 1996).

Main theme: The settling of crystals, and convection currents in the chamber, are the prime causes of layering
'We are convinced that the layered intrusions . . . are due to bodies of homogeneous magma cooling and crystallizing in large chambers, the crystals accumulating on the floors of these chambers to form distinctive layers . . . the action of convection currents appears to be the most likely mechanism to account for most types of fine-scale layering.'
L.R. Wager and G.M. Brown (1968: 546–47), Oxford University

The contribution made by L.R. Wager and G.M. Brown (1968) of Oxford, England, drew on a lifetime of research pursued in the British Isles, Greenland, North America, Africa and other places. In their *Layered Igneous Rocks* the concepts of crystals settling under gravity, of circulating convection currents, and of density currents of heavy, crystal-laden magmas plunging down the cooling sides of the intrusions are inextricably bound together. This became the prevailing doctrine for several decades.

Main theme: Layered igneous rocks simulate the structures of sedimentary strata during crystallisation
'. . . the one principal feature that unites ultramafic rocks of this type is that they are sedimentary in nature – they have sedimentary textures, they are pervasively layered on a grain-by-grain scale, they obey the laws of superposition and they can be recognised and classified in much the same way as ordinary surficial sedimentary rocks.'
E.D. Jackson (1971), United States Geological Survey

Jackson stressed the similarities in structures and textures of layered igneous rocks resulting from crystals settling through molten liquids, and sediments laid down under water.

Main theme: Layering is due to rhythmic supersaturation of the magma
'. . . theory demands that crystallization occurs under supercooled conditions, and that the crystallization of one primocryst phase may change the composition of the magma in such a manner that the nucleation rates of other primocrysts are influenced.' [In somewhat less technical language, this implies rhythmic alternation in the crystallization of different mineral species.]
S. Maaloe (1978), Bergen University, Norway

'The layering in sequences which show excellent size-density sorting . . . is almost certainly due to gravity sorting. However, other layered sequences . . . may be better explained by rhythmic crystallization.'
I.H. Campbell (1978), Australian National University

Significant assaults on the concepts formulated by Wager and Brown were mounted by Sven Maaloe (1978) at the University of Bergen in

Norway and Ian Campbell (1978) at the Australian National University. Their focus was upon the mechanisms governing crystal growth.

Main theme: Layers form by crystal growth on the floors and sides of the chamber
'Layering cannot have been caused by density currents sweeping across the floor . . . it cannot be assumed that crystallization of large mafic intrusions involved crystal settling . . . We propose an alternative origin . . . governed by the relative rates of chemical and thermal diffusion during cooling.'
A.R. McBirney and R.M. Noyes (1979), University of Oregon

McBirney and Noyes were sceptical about the Wager and Brown model. Their reservations were based on the fact that magmas are non-Newtonian liquids. That is, although liquid, they possess a certain strength (the yield strength) that must first be overcome before crystals can drift downwards under the pull of gravity.

Main theme: Changes in confining pressure are fundamental
'It is suggested that changes in total pressure . . . were the prime factor in controlling the changes in mineral assemblages which are represented in the layering . . .'
E.N. Cameron (1980), University of Wisconsin

It was experimental petrology that convinced Eugene Cameron (1980) at the University of Wisconsin, and consultant for Union Carbide Corporation in South Africa, that if the pressures applied to a crystallising magma were to change, the mineral phases separating from it would also change. The appeal in this approach lies in its ability to explain how single layers could develop more or less simultaneously over enormous areas, as in the Bushveld Complex. Valid criticism lies, however, in the question of how pressure changes could be sustained within a magma chamber beneath a thin roof that was floating upon it during intrusion and subsequent crystallisation.

Main theme: Liquids in magma chambers separate into layers from which the rock layers subsequently crystallise
In 1983 '. . . a new mechanism for the solidification of certain kinds

of layered intrusions . . .' was presented, a specific example being 'the extensive layers of the . . . Bushveld . . .' Evidence is offered 'that the Stillwater complex is the product of [the mixing of] two parent liquids'. This resulted in the breaking up of the magma column into a sequence of liquid layers, with crystallisation of these yielding the separate rock layers.
T.N. Irvine et al. (1983), Carnegie Institute of Washington

A new approach was introduced by the study of the waters of the oceans (Huppert & Turner, 1981), and extension of the principles recognised there to magma chambers. The concepts are complex, but may be summarised as the formation of superimposed liquid layers, with each having its own composition, density, and/or temperature. This is known as a double diffusive system, and, as applied to magma chambers by Irvine, Keith and Todd (1983) in the USA, each liquid layer will then crystallise a different rock layer.

Main theme: Layers may form after crystallisation
'Mineral segregation can occur during aging of crystal assemblages as areas composed of nominally larger grains grow at the expense of nearby, smaller grains . . . several features . . . are consistent with at least additional development or enhancement of the layering during aging [of crystal aggregates]'.
A.E. Boudreau (1994), Duke University, USA

A mathematical treatment of the properties of crystal aggregates (Boudreau, 1994) proposes that a layered structure may develop in largely solid rock, after crystallisation, by diffusion of ions through the mass. This model may be valid for fine-scale structures on the scale of millimetres, but probably falls short where layers are metres to tens of metres thick.

Main theme: No single explanation covers all cases
'No single process can account for all forms of layering . . . large-scale magma mixing is the most plausible mechanism . . . to produce chromitite layers. By contrast . . . magnetite layers were . . . produced from relatively thin liquid layers. The origin of incomplete cycles [of layers] . . . appears to be intimately bound up with the emplacement of fresh magma batches [which] mixed with partially crystallized residual liquids . . .'
H.V. Eales and R.G. Cawthorn (1996)

Research sponsored by the CSIR in South Africa during 1985–88 generated a large chemical, isotopic, and field database to test various hypotheses. The compilation by Eales and Cawthorn (1996) concluded that no single process can account for all forms of layering in an intrusion as large as the Bushveld Complex. For example, different mechanisms seem to be required to form layers of chromium ores and iron-titanium ores (composed, respectively, of the minerals chromite and titanomagnetite). Chromium has very limited solubility (<0.2 per cent Cr_2O_3) in basaltic liquids, and to produce a single 1-metre layer of chrome ore a head of liquid at least 500 to 800 metres deep would be needed. By contrast, iron is a major component of the magmas, and relatively small volumes of liquid are capable of supplying all the iron needed to form thick layers of the vanadium-bearing ore, magnetitite.

The above sample of opinions is a rather arbitrary selection that illustrates the separate camps within which the protagonists of competing hypotheses have banded themselves together. Within each camp there are factions who have introduced further subtle variations on the main themes. Furthermore, one or more processes might well have acted at the same time as one or more others, when a good case could be argued for each in turn.

AN AMBITIOUS BUT IMPOSSIBLE EXPERIMENT

Science advances by carrying out well-designed tests, so would it not be possible to simulate, by experiment, a natural process such as this, so as to understand it better? The short answer is 'No'. Consider, for example, the highly improbable scenario where the construction of an experimental model of a relatively small magma chamber several kilometres in radius and 3 kilometres deep is contemplated – in itself only a very small-scale analogue of the Bushveld Complex. The chamber is to be constructed in such a way that it can be filled with rock which is then to be brought up to its melting temperature (around 1 300 °C), under pressures thousands of times that of atmospheric pressure, to generate a liquid magma. This is then to be allowed to cool, and the result examined.

Setting aside the implications of immense effort and prohibitive expense in creating this unlikely laboratory, there will always remain one further parameter that can never be simulated – the passage of time.

The Bushveld magma chamber is estimated to have been filled over a period spanning some 75 000 years, and to have required something like a further 200 000 years for crystallisation as it cooled (Cawthorn & Walraven, 1998: 1685). Any attempt to speed up the processes in the experimental chamber would merely be self-defeating, as this would alter the essential conditions of the experiment. The problem is not unlike the philosophical question about whether the light bulb really does switch off on closing the door of the refrigerator – any trick to confirm that it does would change the nature of the test, and render it invalid. In short, the process can never be rigidly investigated by experiment. It would all be wasted effort.

This impasse leaves only one remaining avenue – analogy. Laws of physical behaviour may be formulated by the study of fluid systems that simulate natural magma chambers – foundries in which metals are cast, tanks in which industrial glass is produced, or by observing the behaviour of water in the oceans and in reservoirs, or the precipitation of salts from solutions in experimental tanks. The literature abounds with accounts of such systems, but in the end it must be accepted that they are merely analogue systems, with no guarantee that the models are appropriate. This is the reason why there are so many hypotheses, but so few 'final answers', in geological science. There will always be uncertainty.

LAYERS TRANSLATE INTO MINERAL WEALTH

It might seem a wild claim to assert that South Africa would be much the poorer if thick bodies of some magmas did not show this perplexing tendency to freeze in layers, but a few examples will show the truth of it. Consider the case of the platinum-group metals. In the Complex, these occur as tiny grains of natural alloys and other minerals in which the precious metals are chemically combined with sulphur, arsenic, tellurium, etc. The highest concentrations occur within two specific rock layers – the Merensky Reef (see Plate 24), and a layer of chromium ore known as the Upper Group 2 (UG2) Chromitite (see Plate 25).

These layers are typically a metre or less in thickness, into which the precious metals were gathered. Each lies at the base of what is termed a Unit, which is commonly 10 to 20 metres thick, or more. A level of ten parts of precious metal per million parts of host rock constitutes a highly profitable grade of ore in the basal layer mined, but if the ore minerals had been uniformly distributed through the full thickness of the Merensky or UG2 Units, the average grade of ore would have been less, even very much less, than one part per million. At this level of dilution, mining and recovery of the precious metals from deep underground would be uneconomic.

The situation with regard to the ore of the base metal, chromium, is much the same. The ore mineral, chromite, contains approximately 45 per cent of chromium oxide (Cr_2O_3) and it forms many layers of varying degrees of purity and thickness that establish this country as the major world source of the metal. Had nature been less bountiful, and disseminated the commodity uniformly throughout its host rocks, their average chromium oxide content would have been a mere 1.4 per cent dispersed through 1 to 2 kilometres of waste rock (Eales, 2000). This would not rate even as a low-grade ore.

WHERE DOES THE CONTROVERSY LEAVE US?

Nature does not lightly surrender its answers. Some would see the layering as predetermined by the development of liquid layers even before crystallisation begins, not unlike the layers of oil, vinegar, and macerated herbs into which a bottle of salad dressing will separate if left undisturbed on a shelf. Others would see layering as a consequence of the way crystals grow and separate from a parent liquid, like crystals attaching themselves to the sides or settling to the base of a jar of honey. Others, again, would attribute it to processes active during or after the final stages of crystallisation.

Whether we fully understand layering or not, the advantages it bestows will be obvious. Without it, South Africa would not have the mineral wealth that it does. The mining of platinum-group metals has been rendered more efficient than extracting them from great tonnages of low-grade ore. As a consequence, market prices have been governed as much by consumer demand as by availability. Similarly, the huge reserves of chromium ore have permitted its recovery and marketing at reasonable prices, as witnessed by its widespread use in chrome plating in the motor industry, stainless or surgical steel, or commonplace appliances in the home – utensils, washing machines, taps and even the kitchen sink.

There is good reason to give profound thanks for the phenomenon of layering!

39

Platinum: From ugly duckling to swan

'. . . the Indians [of the Choco District in South America] regarded the substance as "unripened gold" which they planted back in the river bed to give it more time to ripen into the yellow metal gold . . .
The Spaniards gave the metal the name "platina del Pinto", "platina" being a somewhat derogatory diminutive of "plata" meaning silver . . . [they] took little notice of it beyond throwing it away as a nuisance in their search for gold . . . it occurred in the mines in varying amounts but where it was present in considerable quantities it made working uneconomical and the mine was closed . . .'

D. McDonald (1960: 11)

'That the South African platinum-producing industry . . . has disappointed the more sanguine hopes originally entertained for it is undeniable . . .'

Geological Survey of South Africa (1940: 192)

'The Bushveld Complex probably contains the world's largest Platinum Group Element resources, the monetary value of which is more than three times that of the gold in the entire Witwatersrand basin. At the present rate of production . . . the reserves are sufficient to last more than 12 000 years.'

Dr B. Worst (1991)

The story of platinum is one that has unfolded in much the same way as Hans Andersen's tale about the ugly duckling that grew into a swan – changing scorn and rejection to joyful recognition with all the promise of a splendid future. The historical review by McDonald (1960) and the assessment by Dr Ben Worst, formerly of the Geological Survey of Southern Rhodesia, reveal the complete turnaround in the attitude of the financial and industrial world towards the value of the metal after the nineteenth century.

McDonald recounts that although platinum jewellery of Pre-Inca age had been discovered in Ecuador, the metal was first formally recognised only in 1557 by the Italian scholar, Julius Caesar Scaliger. The source was in alluvial gravels in the Choco district of Colombia, South America. Here, gold was mined during the Spanish occupation, but the platinum occurring there was treated as dross, demanding separation from the gold either by hand-picking or amalgamation of the gold particles with mercury.

So little value was placed on the metal that, McDonald tells us, Dutch traders who had bought gold from Spaniards in the eighteenth century returned by ship to hang the sellers from the yardarm when it was discovered that the yellow metal had been adulterated by alloying it with platinum. Concern about fraud such as this was enough to prompt the Spanish government to prohibit the importation of platinum into Europe, and to decree that it be thrown back into rivers or the sea.

The lack of interest in the metal during the earlier part of the eighteenth century can be attributed to some of its more obvious properties. Like gold, it is insoluble in nitric or sulphuric acid but soluble in aqua regia, so the two could not be separated by simple dissolution. With a melting point of 1 773 °C, it could not be melted by any of the techniques available at that time, although it could be alloyed with metals of lower melting point such as copper, silver, arsenic and antimony. The alloy with gold was hard and brittle, quite unsuitable for shaping or handworking by jewellers. The only practical use for it was in the manufacture of superior mirrors for telescopes.

NEW USES FOR THE METAL

Towards the end of the eighteenth century the unique properties of the metal had created a demand for its use in jewellery and laboratory ware, and for the touch-holes of guns or cannon where the gunpowder was ignited. It was indestructible in fire, free from tarnish in air, brilliant in lustre without gold's yellow tint, but malleable and ductile withal. To illustrate the point, success had been achieved in drawing it into wires as slender as 0.002 centimetres in diameter (McDonald, 1960).

By 1804 it was established that the metal being produced was not a pure element, and four other rare metals within the platinum group – palladium, rhodium, iridium and osmium – were isolated by English chemists Wollaston and Tennant (the remaining member – ruthenium – was isolated only in 1844). Platinum became widely used in the making of boilers for concentrating sulphuric acid to oil of vitriol, and for drawing fine wires for the crosswires of telescopes. Between 1800 and 1821 Wollaston had produced 36 000 ounces from 47 000 ounces of the native metal, and the quality of his product was not surpassed for close on another 100 years (McDonald, 1960). Rhodium was introduced into the manufacture of pen nibs.

The discovery of rich alluvial deposits in Russia in 1824 led the Imperial Russian government to declare a state monopoly, and to issue some half a million ounces as platinum coinage during the period 1828–46. Subsequent withdrawal of the coinage then caused a slump in the importance of the metal, but from the 1860s until 1917 Russia remained the prime world source.

THE EMERGENCE OF THE REGAL SWAN

With time, platinum-group metals became progressively absorbed into the fields of industry, medicine and jewellery. PGM catalysts are used to produce petrol (gasoline) by heating natural petroleum under pressure in the presence of hydrogen. In the automobile industry, the millions of vehicles spewing out noxious gases had led to the introduction of the Clean Air Act in the USA in 1970. The urgent need was for exhaust systems in which the catalyst played a dual role – oxidising poisonous carbon monoxide and unburned fuel to less harmful carbon dioxide, while simultaneously reducing nitrogen oxides to nitrogen. The number of vehicles now in use in densely populated areas across the world underscores the huge potential demand for these metals – 1 to 2 grams of platinum and 0.3 grams of rhodium for each vehicle (Okafor & Coville, 1999). The development of fuel cells for providing power without pollution holds the promise of 10 grams of platinum being needed for each vehicle in the future (Vermaak & Van der Merwe, 2000). Other uses include the manufacture of anti-cancer drugs used in chemotherapy (Neuse, 1999), of ruthenium in corrosion-resistant stainless steels, wear-resistant drilling and cutting tools, and sparkplug electrodes with a life expectancy of 60 000 kilometres or more (Wolff, 1999).

The jewellery trade takes up some 40 per cent of PGM production. By 1997, Japan was using 1.39 million ounces per year, with China being the fastest-growing market, but Japan's later economic woes have led to a decline in consumption (Vermaak & Van der Merwe, 2000: 67). MINTEK (formerly the National Institute for Metallurgy) in South Africa reacted to this by researching platinum alloys tinted to various attractive shades of yellow and pink (Taylor & Biggs, 1999). Where the inherent softness of platinum metal had always been an inhibiting property, it was now shown that improved metallurgical treatment of alloys with copper, cobalt and tungsten can effect very significant hardening. The production of a remarkable alloy of platinum, aluminium and titanium was unveiled – an alloy that is brilliant in lustre and hard enough to be cut and polished into a silver-coloured gemstone.

An indication of the growing demand for platinum metal is to be found in its price. During the mining of gold in South America by the Spanish conquerors platinum was simply thrown away. At the beginning of the nineteenth century Wollaston in England was charging 13 shillings per ounce for the refined product (McDonald, 1960). By 1832 Russian coins were selling for 25 shillings per ounce, and 75 shillings as the century ended. Between 1926 and 1930 the price slumped from £19 per ounce to a third of this. In the 1990s the price lay within the range US$350 to 450. In 2003 it broke the US$700 barrier and moved close to $800, and in 2006 it rose above $1 300 per ounce.

Thus it was that the ugly duckling amongst metals grew into a swan.

SOUTH AFRICA'S PLATINUM TREASURE CHEST

It was against this background that the PGM-mining industry developed in South Africa, although the early days offered little promise of success. The first mention of platinum here was seemingly made in 1906, when samples of chromite ore from the Bushveld Complex were found to contain up to 1.2 pennyweight (around 1.9 g) per ton of ore, but nothing came of this. As late as 1923, the leading economic geologist of the day in South Africa, Dr Percy Wagner, had concluded that 'it would never pay to work the chromite rock for that metal' (Cawthorn, 1999a).

In that same year platinum showed up in a totally different geological setting – in panned samples of soil brought up by termites in the Naboomspruit district (Worst, 1991). The first platinum mine in this country opened here on the Waterberg Lode, but proved to be uneconomic and closed again in 1926. It was in June 1924 that the first meaningful finds were reported. Grains of the metal had been panned from soils in a dry river bed by a prospector, A.F. Lombaard, at Mooihoek in the eastern Transvaal. The geologist, Dr Hans Merensky, was then contacted to direct the search for the primary source. By quite extraordinary coincidence, platinum was within months found in two different host rocks, with different modes of origin and relative ages, less than 4 kilometres apart. In one, the metal occurred as high-grade concentrations in pipe-like bodies that penetrated through the layered sequence of the Bushveld Complex but tapered, carrot-like, with depth. The other discovery was of a distinctive rock layer within the stratified sequence, subsequently named the (now world famous) Merensky Reef (see Plate 24).

The early years following the discoveries were hit hard by the Great Depression, with actual closure of the mines in 1932; nevertheless, by the 1950s South Africa was producing about one-third of the world consumption of around 0.3 million ounces (Cawthorn, 1999b), and by 1998 some 75 per cent of the demand for around 5 million ounces. The output for 2003 rose to 8.5 million ounces (total PGM). Despite Wagner's pessimistic forecast in 1923, the UG2 Chromitite Layer had by now emerged as a PGM reserve some one and a half times greater even than that of the Merensky Reef (see Plate 25).

The reserves are now known to be enormous. The estimates of Vermaak and Van der Merwe (2000) classify 45.7 million ounces PGM + gold as a 'proved reserve', while the overall precious metal in the Complex may be as high as 2.5 billion ounces. Their figure is comparable with that of Jones (1999), 2 billion ounces, or that of 2.25 billion ounces computed by the Department of Mines and Energy (2004). At May 2006 prices, this would equate to an overall value of US$3. 25 trillion!

WHY THE BUSHVELD COMPLEX?

The question: What is the origin of this stupendous bonanza? has exercised the minds of geologists for the last 80 years. Despite years of meticulous research employing the diverse tools of mineralogy, microscopy, geophysics, chemistry and isotope chemistry, the jury is still out on this. In earlier days it was believed that the original Bushveld magmas must have been uniquely enriched in PGMs, but this explanation fell apart when research showed that this was not true, and that the original liquids were in fact little different from common basaltic rocks the Earth over. The parent liquids had held less than 20 parts per billion of platinum, in the order of 1/400 the concentration now found in the orebodies.

The focus therefore shifted to the processes that might have been active in the great mass of liquid that had been injected. The term 'great mass of liquid' is used advisedly here – any intrusion 7 to 9 kilometres thick and extending around 500 kilometres from east to west is worthy of the phrase. Here lies the key. Even minute traces of any component held within a liquid will yield a significant resource if there is a large enough volume of it to start with, provided that nature provides some means of concentrating it within a smaller volume. The geological literature has thrown up a profusion of opinions and hypotheses on the topic. Most of these lie within the domain of physical chemistry, and are not explored here. For those sufficiently intrigued, easily digested reviews are to be found in Viljoen (1999) and Eales (2001).

The threads of controversy are as tangled here as were those in debates on continental drifting in earlier days, or on the origin of our vast gold deposits. R.G. Cawthorn (1999c), an authority in this field at the University of the Witwatersrand, puts the status quo succinctly in the words, 'Viable models which will explain all the complexities associated with platinum reefs, or even represent the dominant accumulation process, remain elusive.'

A first-year university examination answer script once claimed that 'The Upper Critical Zone of the Bushveld Complex is 1 000 metres thick, and consists of solid platinum.' South Africa should be profoundly grateful that this howler does not depict reality, as this would have degraded the price to perhaps a little more than that of iron ore. It is the rarity of the platinum-group metals on a global scale, and their unique chemical and physical properties, which give them their value.

Final words

This book has been written around the twin themes 'consensus amongst the majority at any given time is not proof of anything in Earth science', and 'the dogma of today may well prove to be the heresy of tomorrow'. If any justification of this view is needed, we need only recall how drastically ideas have changed within the last 200 years with respect to what is written in the chapters of Genesis and Exodus. Earth science does not seek to refute the Scriptures, but it does question the perceptions of the men who wrote the texts, and those who then selected what was to be taught. The proofs of the age of the Earth, that the Sun does not revolve around the Earth, or the impossibility of a total inundation of all land as claimed in the legend of Noah's Flood, bear this out.

A great deal of Earth science throws up one controversy after another. It is not a science that enjoys the absolute certainties that chemistry, physics or mathematics do. The experiments of chemistry (for example, that Substance *a* + Substance *b* will consistently react to produce *c*) can be repeated with confidence. The basic principles of physics (in mechanics, optics, electromagnetics, etc.) yield laws that lead to consistent answers. In Earth science there is not the same certainty because, in general, we see only the end-products of processes that were active so far in the past that we have no real feeling for how these processes may have changed through the immensity of time. The events have all too often occurred at depths to which we are unable to penetrate. Direct observation is often impossible, which leaves analogy as the only fruitful line to pursue. It therefore becomes a science of assessing what is possible and what is not, and selecting a model that best explains what we see today. Models evolve and are discarded as new evidence emerges, and in this way there is an edging towards the truth. Could there be any soil more fertile than this for breeding the controversies that characterise the science?

Glossary

albedo the degree of reflecting power of a surface

ammonite extinct marine mollusc of the class that includes surviving forms such as the cuttlefish, squids and nautilus

amygdales vapour cavities (bubbles) in lavas, filled with mineral matter (such as agate)

anorthosite a coarse-grained, generally white or grey rock made of feldspar, often found in layered complexes such as the Bushveld Complex

basalt abundant, dark-coloured, fine-grained rock of volcanic origin

bivalves molluscs with generally calcareous shell of two symmetrical parts joined by a hinge (mussels, clams, oysters)

brachiopods small marine animals enclosed by asymmetric upper and lower shells, and anchored by a stalk

breccia rock composed of angular fragments cemented together by fine-grained matrix material

chert hard, flint-like, ultra-fine-grained rock, composed of silica (SiO_2)

coelenterates large phylum of animals including corals, sea anemones and jellyfish

conglomerate sediment of rounded pebbles or boulders cemented together by a finer-grained matrix

convection bodily movement within a viscous medium due to differences in temperature or density

crinoids marine animals closely related to sea urchins (echinoids) with flexible arms and a stem

crust of the Earth the outermost shell, a few tens of kilometres in thickness, enclosing the mantle

density the ratio of mass to volume, commonly expressed as grams per cubic centimetre (gm/cm^3)

diabase a term commonly applied to highly altered (degraded) basaltic rocks

dolomite sedimentary rock composed of calcium-magnesium carbonate

dyke narrow, vertical or near-vertical, wall-like body of intrusive igneous rock

echinoids sea urchins – marine animals of globular or discoidal shape, with a shell of calcareous plates covered with spines

electrum natural alloy of gold with approximately 50 per cent silver, pale yellow in colour

eluvial genetic term for residual deposits left behind after removal of the decomposed rock material

facies term applied to different parts of bodies of rock that exhibit a change in physical or chemical attributes

feldspars the dominant group of minerals in the Earth's crust; they are transparent or translucent (commonly white or pink) silicates of potassium (K), sodium (Na) and/or calcium (Ca)

felsite extremely fine-grained volcanic material with a composition similar to that of granite

fluvial a term pertaining to the action of rivers

foraminifera Protozoa, the lowest forms of animals; as marine fossils, occurring as single- or many-chambered calcareous shells, often of pinhead size

gabbro the coarser-grained (more slowly cooled) equivalent of basalt

gastropods molluscs with a spiral external shell (snails, cowries, whelks) or no shell (slugs)

geochemistry the study of the chemical composition of rocks and minerals, and what can be inferred from the data

geochronology the science of dating of rocks

geothermal gradient the rate of increase of temperature with depth within the Earth

gneiss coarsely banded metamorphic rock, often with a composition approximating to granite

granite abundant, coarse-grained, igneous rock composed of feldspars and quartz, emplaced at depth

graptolites extinct, small, colonial, marine animals confined to Lower Palaeozoic sedimentary rocks

guyot seamount (volcanic peak) rising from the sea floor, having a flat top below sea level

igneous rocks rocks such as granite, syenite, basalt, etc., that crystallise at high temperature from largely molten liquids (magmas)

intrusive general term for a body of rock formed by injection of wholly or partially molten liquid (magma) into older rocks

invertebrates animals without a backbone

isotope atomic species with the same atomic number (number of electrons or protons) but different masses

kimberlite igneous rock forming the matrix in which diamonds may be found

lahar destructive torrent of mud and debris, often associated with volcanic eruptions

limestone rock composed of calcium carbonate

lithification conversion of unconsolidated material into rock

littoral pertaining to the seashore

mafic rocks igneous rocks characterised by a high proportion of dark, heavy minerals

magma high-temperature melt with or without some small proportion of crystals

mantle the material that extends from the base of the crust to the metallic core of the Earth

marble limestone recrystallised by metamorphism to a dense, even-grained rock

metamorphic rocks rocks with characteristic structure, texture and/or minerals, produced by heat and/or pressure acting upon some pre-existing rock (gneiss, schist, marble, slate, quartzite)

mineral a naturally occurring crystalline material defined by its chemical composition, properties, crystal structure and mode of occurrence

moraine rock debris deposited by glaciers

olivine a green or brown silicate of magnesium and iron, $(Mg, Fe)_2 SiO_4$, found in SiO_2-poor rocks

ore rock bearing some naturally occurring mineral (diamond, gold, iron oxide, etc.), that can profitably be mined

palaeontology the study of fossils and the information they yield

petrology the study of the textures, composition, origin and classification of all rocks

pyrite a common, brass-coloured, hard mineral in which iron and sulphur are chemically combined

pyroclastic hot ash, lava fragments and other debris ejected during explosive volcanism

pyroxene group of silicate minerals including orthopyroxenes such as bronzite $[(Mg,Fe)SiO_3]$ and clinopyroxenes – such as diopside $[Ca(Mg,Fe)Si_2O_6]$; common species are grey, black, brown or green

quartz hard, transparent mineral (SiO_2) resembling glass, which may also occur in well-formed hexagonal crystals; one of the most common minerals of the Earth's crust

quartzite hard, compact rock composed dominantly of quartz, formed by recrystallisation of sandstones by heat and/or pressure

radioactivity process whereby certain unstable atomic nuclei decay to form different atomic species, with the emission of energy

radiolaria tiny marine Protozoans with a skeleton of SiO_2, often forming deep-sea oozes on the ocean floors

refractive index a fundamental optical property governing the refraction (bending of the path) of light waves entering a medium, and the reflectivity of its surfaces

rhyolite extremely fine-grained volcanic material with a composition similar to that of granite

roches moutonnées rounded, hummocky outcrops of rock shaped by glacial abrasion

schist metamorphic rock in which platy, flaky or elongated crystals impart a foliated character

seamount volcanic peak rising from the sea floor

sedimentary rocks rocks deposited by the agents water, wind or ice, or by explosive volcanism, or consisting of animal or plant remains

seismology the study of earthquake waves transmitted through the Earth

slate a fine-grained metamorphic rock that splits almost perfectly along natural planes of cleavage

specific gravity the factor by which something is heavier or lighter than the same volume of water

spinel minerals in which iron, magnesium, aluminium, chromium, etc., are combined with oxygen

stoping enlargement of a cavity or chamber by the breaking off of blocks from its walls

stratigraphy the study of the sequence of rocks, in terms of their age and composition, and the correlation of them in different regions or continents

terrane a series or sequence of rocks defining the geological character of a particular region

tillite unstratified rock resulting from consolidation of the debris (till) left behind by glacial ice

trilobites extinct sub-class of crustaceans with some resemblance to crabs

xenolith fragment of some rock foreign to that in which it is enclosed

zeolites group of minerals commonly found filling vesicles (bubble cavities) in lavas

NOTES

1. The date on which the first vertebrate fossils were discovered in Karoo strata is not entirely clear. 'Reminiscences of Bain' (1856) reprinted in the *Transactions of the Geological Society of South Africa* for 1895 describes how he made his first collection of bones one mile north of Fort Beaufort. The date given is 1883 – an obvious typographic error. It was shortly after this, Bain continues, that he unearthed the 'bidental' skull that gave rise to the genus *Dicynodon*, subsequently so named by Professor Owen at the British Museum. The despatch of this collection to the Geological Society of London is acknowledged in a letter of thanks from the president of the society, Henry Warburton, together with an award of twenty guineas. This is dated January 1845, suggesting that the probable date of the first discovery was 1838 – misprinted as 1883.
2. As an illustration of this, the evaporation of the Mediterranean Sea alone, between 7.2 and 5.4 million years ago, yielded deposits of evaporites holding 'one-sixth of the salinity of the World' (Cita, 2001: 354). The story of the drying up of the Mediterranean is briefly told, for example, in the engaging book, *Noah's Flood*, by Ryan and Pitman (2000). The isolation of the Mediterranean Sea from the North Atlantic had been the result of continental drifting, with Morocco in North Africa ramming into Spain to form a barrier at Gibraltar. The drying up of the Mediterranean Sea that followed was extreme, leading to the deposition of evaporites (the minerals selenite, anhydrite and halite, or salt) on the floor of the basin. Ultimately

the region had become a desert-like depression before the breaching of the Gibraltar 'dam' had once more flooded the basin with sea water.

3. To illustrate the point, we might cite an extreme example. The base of our Karoo Sequence (around 320 million years in age) in South Africa may rest directly in places on old Basement rocks in the order of 3.5 billion years in age, leaving, locally, an unrecorded lacuna or gap of billions of years.

   Taking the thickness of the sequence of rocks constituting the South African Witwatersrand Supergroup as a maximum of 24 000 feet (7.3 km), and the lapse of time between the deposition of the base and top of the sequence as 210 million years (Hartzer et al., 1998), the average rate of accumulation would have been *~0.011 foot (0.33 cm) per century*. Alternatively, we might use the bracketing by England et al. (2002) of 7.5 kilometres of sediments between the underlying Dominion Group (3 074 million years old) and overlying Ventersdorp Supergroup lavas (2 714 million years old) to derive a figure of 0.208 cm/century. The overlying 54 kilometres of the Pretoria Group would also yield, by the same calculation, 0.213 cm/century.
4. It was even suggested by the sedimentologist, F.J. Pettijohn of Johns Hopkins University (1957) from the few absolute-age determinations available at that time, that 'averages' are time-dependent. Figures he offered suggested 3.35 centimetres per century for the Cenozoic, 2.74 centimetres per century for the Mesozoic, 2.13 centimetres per century for Late Palaeozoic and 1.52 centimetres per century for Early Palaeozoic. Other more recent estimates range from 0.005 centimetres per century for deep-sea clays and oozes (Bayly, 1968) to 3.66 centimetres per century for sediments of the continental shelves (Eardley, 1972).
5. Hence, quartzites (originally sands) were held to have been changed into norites (the most common igneous rock-type of the Complex), shales (originally muds) into pyroxenites, limestones into layers of magnetic iron ore, calcareous cherts into anorthosites composed simply of feldspar, and other calcareous sediments into layers of chromitite (chromium ore) and platinum-bearing horizons.
6. There are significant differences of opinion concerning the talent, the ancient measure of weight. Where Hoover and Hoover adopted a conversion of 1 talent = 837 troy ounces, Funk (1903) gives it as 27.21 kilograms or 875 troy ounces for the ancient Greek talent, but nearly twice that amount (50.12 kg or 1 612 troy ounces) for the Hebrew talent. Andrews's (1999: 77) conversion of the 120 talents presented by the Queen of Sheba as a meagre 6 kilograms is out of kilter, indicating either a misprint or arithmetic slip. Using Funk's conversion factor of 50 kilograms, the gift equates to 6 014 kilograms (not 6 kg). Even using Andrews's value (1999: 307) of 1 talent = 27 pounds (not kg),

Sheba's gift would have been a substantial 1 472 kilograms of gold.

7. One troy ounce equals 31.1 grams, or 20 pennyweight (dwt), and one pennyweight is 1.5552 grams. Assay values are commonly quoted as dwt per ton, or grams per ton; 1 gram per metric tonne equals 0.583 dwt per short ton of 2 000 pounds, or 0.643 dwt per metric tonne of 2 204.6 pounds.
8. Perhaps Young's most telling point, as a convinced infiltrationist, was his stressing of 'the intimate association of gold and pyrite . . . one of the most conspicuous features of epigenetic gold occurrences [deposits produced by invading solutions]' (1911: xxix). He gave short shrift to the assertion that particles of carbonaceous matter that are dispersed in the ores (which he identified as anthracite) had been precipitants of the precious metal.
9. Some 50 to 75 per cent of the total pyrite population is composed of rounded grains, and these are associated with other common detrital minerals such as chromite and zircon. Abraded, often fractured detrital pyrite grains still displaying original dendritic, banded or concentric growth zones were found to be enclosed by younger overgrowths. Variations in trace element and sulphur-isotope characteristics of the crystal cores point to their having been collected from diverse and probably distant sources. Grain-to-grain, place-to-place, and mine-to-mine heterogeneity in composition is markedly greater in the crystal cores than in the overgrowths. The greater homogeneity of composition in the overgrowths points to their having been added while resident in a more uniform environment after transportation to their final resting place – the place where they are now found. In other words, the pyrite grains have grown in two stages, first in some distant source from which they were transported, and then they were coated by subsequent enlargement of the grains after deposition in the gold-bearing reefs.

References

While page numbers are specified in the main text for extracts taken from books, these have mostly been omitted in the case of extracts from shorter papers in journals.

Agricola, G. (1556). *De Re Metallica*. Translated from the first Latin edition by H.C. Hoover and L.H. Hoover (1912, reprinted 1950). Dover Publications, New York, 638pp.

Allen, A. (1997). Thulamela, citadel of hope. *Goldfields of South Africa Annual Review, 1996–97.*

Allen, G. and D. Allen (1978). *The Guns of Sacramento*. Robin Garton Ltd, London, 81pp.

Andrews, R. (1999). *Blood on the Mountain*. Orion Books, London, 346pp.

Antrobus, E.S.A. (1986). *Witwatersrand gold*. Geological Society of South Africa, Johannesburg, 298pp.

Armstrong, H.C. (1938). *Grey Steel: J.C. Smuts*. Arthur Barker, Ltd, London, 406pp.

Armstrong, J.A.H. (1907). Sedimentary deposits in Natal. *Fifth Annual Report of the South African Association for the Advancement of Science*, 71–75.

Asbestos Institute (1986). *Does the EPA really have a case against chrysotile asbestos?* Asbestos Institute, Montreal, Canada, 48pp.

Atherstone, W.G. (1895). Kimberley and its diamonds. *Transactions of the Geological Society of South Africa*, 1, 76–82.

Atherstone, W.G. and T. Bain [*sic*] (1871). Report of a tour in connection with reported gold discoveries in the Gouph, Division of Prince Albert. *Report of the House of Assembly, Cape of Good Hope*, 4pp.

Ayala-Carcedo, F.J., A. Perejón, L. Jordá and O. Puche (2005). The XIV International Geological Congress of 1926 in Spain. *Episodes: Journal of International Geoscience*, 28–1, 42–47.

Bain, A.G. (1856). On the geology of South Africa. *Transactions of the Geological Society of London*, 7, 175–92.

—. (1895). Reminiscences and anecdotes connected with the history of geology in South Africa. *Transactions of the Geological Society of South Africa*, 2, 59–75.

Ballard, R.D. (2001). Black Sea mysteries. *National Geographic Magazine*, 199, 52–69.

Barnicoat, A.C., I.H.C. Henderson, R.J. Knipe, B.W.D. Yardley, R.W. Napier, N.P.C. Fox, A.K. Kenyon, D.J. Muntingh, D. Strydom, K.S. Winkler, S.R. Lawrence and C. Cornford (1997). Hydrothermal gold mineralization in the Witwatersrand basin. *Nature*, 386, 820–24.

Bartlett, J. (undated). *Familiar quotations*. Routledge and Kegan Paul Ltd, London, 524pp.

Barton, J.M., B. Ryan and R.E.P. Fripp (1983). Rb-Sr and U-Th-Pb isotopic studies of the Sand River Gneisses, Central Zone, Limpopo Mobile Belt, 9–18. In W.J. van Biljon and J.H. Legg (Eds.), *The Limpopo Belt*. Special Publication of the Geological Society of South Africa, Johannesburg, 8, 203pp.

Bascom, W. (1961). *A hole in the bottom of the sea: The story of the Mohole project*. Doubleday and Co., New York, 352pp.

Baumann, M. (1919). Ancient tin mines of the Transvaal. *Journal of the Chemical, Metallurgical and Mining Society of South Africa*, 120–31.

Bauval, R. and G. Hancock (1996). *The keeper of Genesis*. William Heinemann Ltd, London, 362pp.

Bayly, B. (1968). *Introduction to petrology*. Prentice-Hall, New Jersey, 371pp.

Bazin, G. (1958). *A concise history of art*. Thames and Hudson, London, 275pp.

Becker, G.F. (1897). The Witwatersrand banket. *Eighteenth Annual Report of the United States Geological Survey*, Part v, 153.

Benton, M.J. (1995). Diversification and extinction in the history of life. *Science*, 268, 52–58.

Berkner, L.V. and L.V. Marshall (1972). Oxygen and evolution, 143–49. In I.G. Gass, P.J. Smith and R.C.L. Wilson (Eds.), *Understanding the Earth*. Artemis Press, Sussex, 383pp.

Bernhard, F.O. (1971). *Karl Mauch: African explorer*. C. Struik Ltd, Cape Town, 247pp.

Bolles, E.B. (2000). *Galileo's commandment*. Abacus, London, 485pp.

Bonney, T.G. (1907). On the supposed kimberlite magma and eclogite concretions. *Transactions of the Geological Society of South Africa*, 10, 95–100.

Booker, C. (2003). The Asbestos Dossier: How can the HSE still tolerate this racket? *The Weekly Telegraph*, London, 26 February 2003.

Borchers, R. (1964). Exploration of the Witwatersrand System and its extensions, 1–23. In S.H. Haughton (Ed.), *The geology of some ore deposits in Southern Africa*. Geological Society of South Africa, Johannesburg, Vol.1, 625pp.

Bordy, E.M., A.J. Bumby, O. Catuneanu and P.G. Eriksson (2004). Advanced Early Jurassic termite nests: Evidence from the Clarens Formation in the Tuli Basin, Southern Africa. *Palaios*, 19, 68–78.

Boudreau, A.E. (1994). Mineral segregation during crystal aging in two-crystal, two-component systems. *South African Journal of Geology*, 97, 473–85.

Bowen, N.L. (1928). *The evolution of the igneous rocks*. Princeton University Press, New Jersey, 332pp.

Brain, C.K. (1995). Back to beginnings. *Phoenix Magazine*. Albany Museum, Grahamstown, 8, 7–9.

—. (1997). Importance of Nama Group sediments and fossils to the debate about animal origins. *Palaeontontologia Africana*, 34, 1–13.

Brewer, E.C. (undated). *Dictionary of phrase and fable*. Odhams Press, London, 1158pp.

Brooks, H. (1985). Technology-related catastrophes: Myth and reality, 109–36. In S. Friedländer, G. Holton, L. Marx and E. Skolnikoff (Eds.), *Visions of Apocalypse: End or rebirth?* Holmes and Meier Publishers Ltd, London, 268pp.

Browning, D.C. (1951). *Everyman's dictionary of quotations and proverbs*. J.M. Dent and Sons, London, 766pp.

Buchanan, P.C. and W.U. Reimold (1998). Studies of the Rooiberg Group, Bushveld Complex. *Earth and Planetary. Science Letters*, 155, 149–165.

Bulpin, T.V. (1955). *Storm over the Transvaal*. Howard Timmins, Cape Town, 319pp.

Caldecott, W.A. (1903). The development of gold extraction methods on the Witwatersrand. *First Annual Report of the South African Association for the Advancement of Science*, 240–49.

Cameron, E.N. (1980). Evolution of the lower Critical Zone, Eastern Bushveld Complex. *Economic Geology*, 75, 845–71.

Campbell, I.H. (1978). Some problems with the cumulus theory. *Lithos*, 11, 311–23.

Carson, R. (1951). The long snowfall, 439–46. In E.B. Bolles (Ed.), *Galileo's commandment*. Abacus, London, 485pp.

Cawthorn, R.G. (1999a). The discovery of the platiniferous Merensky Reef in 1924. *South African Journal of Geology*, 102, 178–83.

—. (1999b). The platinum and palladium resources of the Bushveld Complex. *South African Journal of Science*, 95, 481–89.

—. (1999c). Platinum-group element mineralization in the Bushveld Complex. *South African Journal of Geology*, 102, 268–81.

Cawthorn, R.G. and F. Walraven (1998). Emplacement and crystallization time for the Bushveld Complex. *Journal of Petrology*, 39, 1669–87.

Cita, M.B. (2001). The Messinian salinity crisis in the Mediterranean, 353–60. In U. Briegel and W. Xiao (Eds.), *Paradoxes in geology*. Elsevier Science B.V., Amsterdam, 467pp.

Cody, G.D. (2005). Geochemical connections to primitive metabolism. *Elements*, 1, 139–44.

Coe, A.L., D.W.J. Bosence, K.D. Church, S.S. Flint, J.A. Howell and R.C.L. Wilson (2003). *The sedimentary record of sea-level change*. The Open University, Cambridge University Press, Cambridge, 286pp.

Coertze, F.J. (1958). Intrusive relationships and ore-deposits in the western part of the Bushveld Igneous Complex. *Transactions of the Geological Society of South Africa*, 61, 387–92.

Cohen, J.M. and M.J. Cohen (1963). *The Penguin dictionary of quotations*. Penguin Books, London, 664pp.

Cornell, F. (1920). *The glamour of prospecting*. T. Fischer Unwin, London. (Facsimile Edition David Philip, Cape Town, 1986), 336pp.

Corstorphine, G.S. (1899). Geologist's Report for 1899. *Annual Report of the Cape Geological Commission*, 25–26.

—. (1904). The history of stratigraphical investigation in South Africa. *Second Annual Report of the South African Association for the Advancement of Science*, 2, 145–81.

—. (1907). The occurrence in kimberlite of garnet-pyroxene nodules carrying diamonds. *Transactions of the Geological Society of South Africa*, 10, 65–69.

—. (1911). Note on a diamontiferous garnet-pyroxene nodule from the Roberts Victor Mine, O.F.S. *Transactions of the Geological Society of South Africa*, 14, 71–73.

Courtillot, V. and Y. Gaudemer (1996). Effects of mass extinctions on biodiversity. *Nature*, 381, 146–48.

Cousins, C.A. (1958). Discussion on paper by F.J. Coertze. *Transactions of the Geological Society of South Africa*, 61, 393–94.

—. (1959). The structure of the mafic portion of the Bushveld Igneous Complex. *Transactions of the Geological Society of South Africa*, 62, 179–202.

—. (1970). Contributed discussion on paper by Hamilton (1970), 374–75. In D.J.L. Visser and G. von Gruenewaldt (Eds.), *Symposium on the Bushveld Igneous Complex and other layered intrusions*. Special Publication of the Geological Society of South Africa, Johannesburg, 763pp.

Crane, N. (2003). *Mercator: The man who mapped the planet*. Weidenfeld and Nicholson, London, 397pp.

Crookes, Sir William (1897a). The diamond mines of Kimberley. *Nature*, 55, 519–32.

—. (1897b). Diamonds. *Nature*, 56, 325–31.

Cutler, A. (2003). *The seashell on the mountain top*. Penguin, London, 228pp.

Daly, R.A. (1949). Discussion on paper by S. van Biljon. *Transactions of the Geological Society of South Africa*, 52, 177–78.

Dana, E.S. (1904). *The System of Mineralogy of James Dwight Dana*. John Wiley and Sons, London, 1134pp., with appendices.

Darwin, C. (1859). *On the origin of species*. Ward Lock and Co., London, 1911 Reprinting, 389pp.

Davidson, C.F. (1965). The mode of origin of banket orebodies. *Transactions of the Institution of Mining and Metallurgy*, 74, 319–38.

Da Vinci, L. (*c*.1480). Seashells in the mountains, 105–11. Reprinted in E.B. Bolles (2000, Ed.), *Galileo's commandment*. Abacus, London, 485pp.

Dawson, J. (2003). MIT study sees nuclear power as green weapon against global warming. *Physics Today*, 56, 12, 34–35.

Dawson, J.W. (1880). *The chain of life in geological time*. The Religious Tract Society, London, 272pp.

De Fontbrune, J-C. (1983). *Nostradamus 1: Countdown to Apocalypse*. Pan Books, London, 453pp.

De Klerk, W.J. (1995a). Extracts from Atherstone (1857 and 1871). *Phoenix Magazine*. Albany Museum, Grahamstown, 8, 2 and 3, 28–29.

—. (1995b). The naming of *Paranthodon*. *Phoenix Magazine*. Albany Museum, Grahamstown, 8, 2 and 3, 30–33.

—. (2002). A dicynodont trackway from the Cistecephalus Assemblage Zone in the Karoo, east of Graaff-Reinet, South Africa. *Palaeontologia Africana*, 38, 73–91.

De Kock, M. (1994). Power and wealth were wrought with gold. *Gold Fields of South Africa Annual Review for 1993–1994*, 18–23.

Department of Mines and Energy (2004). *South Africa's mineral industry*. Director of Mineral Economics, Pretoria, 216pp.

De Ronde, C.E.J. and T.W. Ebbesen (1996). 3.2 b.y. of organic compound formation near sea-floor hot springs. *Geology*, 24, 791–94.

De Wit, M.J., R. Hart, A. Martin and P. Abbott (1982). Archean abiogenic and probable biogenic structures. *Economic Geology*, 77, 1783–801.

Dickin, A.P. (2002). *On a faraway day*. Brentwood Christian Press, Columbus, Georgia, 336pp.

Dornan, Revd S.S. (1915). Rhodesian ruins and native tradition. *Thirteenth Annual Report of the South African Association for the Advancement of Science*, 502–16.

Draper, D. (1895). The Dwyka Conglomerate. *Transactions of the Geological Society of South Africa*, 1, 90–112.

—. (1897). A ramble through the geology of Southern Africa. *Transactions of the Geological Society of South Africa*, 3, 19–30.

Dundes, A. (Ed.) (1988). *The Flood myth*. University of California Press, Berkeley, 452pp.

Du Toit, A.L. (1921). Land connections between the other continents and South Africa in the past. *Nineteenth Annual Report of the South African Association for the Advancement of Science*, 120–40.

—. (1925). Report of the Kalahari reconnaissance of 1925. *Report of the Department of Irrigation*, Pretoria, 60, 69pp.

—. (1927a). *A geological comparison of South America with South Africa*. Carnegie Institute, Washington, 158pp.

—. (1927b). The Kalahari and some of its problems. *South African Journal of Science*, 24, 88–101.

—. (1928). The origin of the Dwyka conglomerate. *Geological Magazine*, London, 65, 240.

—. (1937). *Our wandering continents*. Oliver and Boyd, London, 366pp.

—. (1945). Origin of the amphibole asbestos deposits of South Africa. *Transactions of the Geological Society of South Africa*, 48, 161–206.

—. (1954). *The geology of South Africa*. 3rd Edition, S.H. Haughton (Ed.). Oliver and Boyd, London, 611pp.

Eales, H.V. (1960). *Gold fineness in hydrothermal ores*. Ph.D. dissertation, Rhodes University, Grahamstown, South Africa.

—. (1967). Reflectivity of gold and gold-silver alloys. *Economic Geology*, 62, 412–20.

—. (2000). Implications of the chromium budget in the Western Limb of the Bushveld Complex. *South African Journal of Geology*, 103, 141–50.

—. (2001). *A first introduction to the geology of the Bushveld Complex*. Popular Geoscience Series, Council for Geoscience, Pretoria, 2, 84pp.

—. (2003). Caveats in defining the magmas parental to the mafic rocks of the Bushveld Complex. *Mineralogical Magazine*, 66, 815–32.

Eales, H.V. and R.G. Cawthorn (1996). The Bushveld Complex, 181–230. In R.G. Cawthorn (Ed.), *Layered intrusions*. Elsevier Science B.V., Amsterdam, 531pp.

Eales, S.J. (1979). Medieval lapidary lore. *Bulletin of the Geological Society of South Africa*, 22, 13.

Eardley, A.J. (1972). *Science of the Earth*. Harper and Row, New York, 468pp.

Ehlers, D.L. and C.J. Vorster (1998). Asbestos, 68–75. In M.G.C. Wilson and C.R. Anhaeusser (Eds.), *Handbook 16: The mineral resources of South Africa: Sixth Edition*. Council for Geoscience, Pretoria, 740pp.

England, G.L., B. Rasmussen, B. Krapez and D.I. Groves (2002). Palaeoenvironmental significance of rounded pyrite in siliciclastic sequences of the Late Archaean Witwatersrand Basin. *Sedimentology*, 49, 1133–56.

Evans, J. and M.S. Serjeantson (1960). *English medieval lapidaries*. Oxford University Press.

Exton, H. (1896). Dr. Exton on Atherstone's paper. *Transactions of the Geological Society of South Africa*, 1, 86–89.

—. (1897). Presidential address. *Transactions of the Geological Society of South Africa*, 3, 4–8.

—. (1898). Presidential address. *Transactions of the Geological Society of South Africa*, 4, 6–10.

Fermor, Sir Lewis (1938). *Garnets and their role in nature*. Special Publication of the Indian Association for the Cultivation of Science, Calcutta, 6.

Ferris, J.P. (2005). Mineral catalysis and prebiotic synthesis. *Elements*, 1, 145–50.

Fivaz, C.E. (1988). How the MacArthur-Forrest cyanidation process ensured South Africa's golden future. *Journal of the South African Institution of Mining and Metallurgy*, 88, 309–18.

Ford, W.E. (1918). The growth of mineralogy from 1818 to 1918. *American Journal of Science*, 46, 240–54.

—. (1945). *A text book of mineralogy*. John Wiley and Sons, New York, 851pp.

Fortey, R. (2000). *Trilobite!* Harper Collins Publishers, London, 269pp.

Frakes, L.A. (1979). *Climates throughout geologic time*. Elsevier, Amsterdam, 810pp.

Frazer, J.G. (1918). The Great Flood, 113–23. Reprinted in A. Dundes (1988, Ed.), *The Flood myth*. University of California Press, Berkeley, 452pp.

Friedländer, S. (1985). Themes of decline and end in nineteenth-century Western imagination, 61–83. In S. Friedländer, G. Holton, L. Marx and E. Skolnikoff (Eds.). *Visions of Apocalypse: End or rebirth?* Holmes and Meier Publishers Ltd, London, 268pp.

Friedländer, S., G. Holton, L. Marx and E. Skolnikoff (Eds.). (1985). *Visions of Apocalypse: End or rebirth?* Holmes and Meier Publishers Ltd, London, 268pp.

Frimmel, H.E. and W.E.L. Minter (2002). Recent developments concerning the geological history of the Witwatersrand gold deposits, South Africa. Special Publication of the Society of Economic Geologists, 9, 17–45.

Funk, I.K. (1903). *A standard dictionary of the English language*. Funk and Wagnalls Co., New York.

Funkenstein, A. (1985). A schedule for the end of the world, 44–60. In S. Friedländer, G. Holton, L. Marx and E. Skolnikoff (Eds.), *Visions of Apocalypse: End or rebirth?* London: Holmes and Meier Publishers Ltd, 268pp.

Geikie, Sir Archibald (1903). *Text-book of geology*. Macmillan and Co Ltd, London, 1472pp.

—. (1905). *The founders of geology*. University Press, Glasgow, 486pp.

Geological Survey of South Africa (1940). *Mineral resources of the Union of South Africa: Third edition*. Government Printer, Pretoria, 544pp.

Gess, R.W. (1995). Taking the lid off a geological plant press. *Phoenix Magazine*. Albany Museum, Grahamstown, 8, 18–22.

—. (2002). *The palaeocology of a coastal lagoon*. M.Sc. thesis, University of Fort Hare, 67pp and annexures.

Gess, R.W. and N. Hiller (1995). A preliminary catalogue of fossil algal, plant, arthropod and fish remains. *Annals of the Cape Provincial Museums*, 19, Part V, 225–305.

Gevers, T.W. (1949). The life and work of Dr. Alex L. du Toit. *Proceedings of the Geological Society of South Africa*, Annexure to Vol.LII , 1–103.

Gibson, R.L. and W.U. Reimold (2001). *The Vredefort impact structure*. Council for Geoscience, Pretoria, Memoir 92, 111pp.

Glatzmaier, G.A., R.S. Coe, L. Hongre and P.H. Roberts (1999). The role of the Earth's mantle in controlling the frequency of geomagnetic reversals. *Nature*, 401, 885–90.

Gould, Stephen J. (1978). *Ever since Darwin*. Pelican Books, London, 285pp.

—. (1982). Creationism: Genesis vs. Geology, 427–37. Reprinted in A. Dundes (1988, Ed.), *The Flood myth*. University of California Press, Berkeley, 452pp.

—. (1989). *Wonderful life*. Century Hutchison, South Africa, 347pp.

Graton, L.C. (1930). *Hydrothermal origin of the Rand gold deposits*. Economic Geology Publishing Company, Lancaster, Penna., 185pp.

Gray, J. (1937). *Payable gold*. Central News Agency, Johannesburg, 286pp.

Green, A.H. (1888). A contribution to the geology and physical geography of the Cape Colony. *Quarterly Journal of the Geological Society of London*, 44, 239–69.

Green, L.G. (undated, *c.*1950). *To the River's End.* Howard Timmins, Cape Town, 208pp.

Gregory, J.R. (1868a). Diamonds from the Cape of Good Hope. *Geological Magazine*, 5, 558–61.

—. (1868b). On the gold-fields (?) of South Africa. *Geological Magazine*, 5, 561–63.

Gribbin, J. (1975). Dusty embrace of the Milky Way may explain ice ages and the origin of man. *New Scientist*, 65, 695.

Gumbricht, T., T.S. McCarthy and C.L. Merry (2001). The topography of the Okavango Delta, Botswana. *South African Journal of Geology*, 104, 243–64.

Habel, N.C. (1971). The two Flood stories in Genesis, 13–28. Reprinted in A. Dundes (1988, Ed.), *The Flood myth*. University of California Press, Berkeley, 452pp.

Hall, A.L. (1912). *Geology of the Murchison Range*. Memoir 6, Geological Survey of South Africa, Pretoria, 186 pp.

—. (1932). *The Bushveld Igneous Complex of the Central Transvaal*. Memoir 28, Geological Survey of South Africa, Pretoria, 560pp.

Hall, R.N. (1903). Great Zimbabwe. *First Annual Report of the South African Association for the Advancement of Science*, 1, 504–15.

Hallam, A. (1989). *Great geological controversies*. Oxford University Press, Oxford, 244pp.

—. (1992). *Phanerozoic sea-level changes*. Columbia University Press, New York, 266pp.

Hallbauer, D.K. (1983). Characterization of quartz and pyrite in Witwatersrand gold-bearing conglomerates, 39–57. In J.P.R. de Villiers and P.A. Cawthorn (Eds.), *ICAM 81: Proceedings of the First International Congress on Applied Mineralogy*. Special Publication of the Geological Society of South Africa, Johannesburg, 7, 509pp.

—. (1986). Mineralogy and geochemistry of Witwatersrand pyrite, gold, uranium and carbonaceous matter, 731–52. In C.R. Anhaeusser and S. Maske (Eds.), *Mineral deposits of Southern Africa*. Geological Society of South Africa, Johannesburg, Vol.1, 1020pp.

Halpern, P. (1998). *Countdown to apocalypse*. Perseus Publishing, Cambridge, Mass., 290pp.

Hamilton, W. (1970). Bushveld Complex: Product of impacts?, 367–79. In D.J.L. Visser and G. von Gruenewaldt (Eds.), *Symposium on the Bushveld Igneous Complex*. Special Publication of the Geological Society of South Africa, Johannesburg, 1, 763pp.

Hancock, G. (1995). *Fingerprints of the gods*. William Heinemann, London, 578pp.

Hapgood, C.H. and J.H. Campbell (1959). *Earth's shifting crust*. Museum Press, London, 438pp.

Harger, H.S. (1905). The diamond pipes and fissures of South Africa. *Transactions of the Geological Society of South Africa*, 8, 110–34.

—. (1909). Occurrence of diamonds in Dwyka conglomerate and amygdaloidal lavas. *Transactions of the Geological Society of South Africa*, 12, 139–58.

—. (1945). The early history of the Geological Society of South Africa. *Transactions of the Geological Society of South Africa*, 48, i–iv.

Harte, J. (2002). Towards a synthesis of the Newtonian and Darwinian worldviews. *Physics Today*, October 2002, 29–34.

Hartzer, F.J., M.R. Johnson and B.M. Eglington (1998). *Stratigraphic table of South Africa*. Council for Geoscience, Pretoria.

Hatch, F.H. and G.S. Corstorphine (1905). *The Geology of South Africa*. Macmillan and Co., New York, 348pp.

Hatton, C.J. and J.K. Schweitzer (1995). Evidence for synchronous extrusive and intrusive Bushveld magmatism. *Journal of African Earth Sciences*, 21, 579–94.

Hawthorne, J.B. (1975). Model of a kimberlite pipe, 1–15. In L.H. Ahrens, J.B. Dawson, A.R. Duncan and A.J. Erlank (Eds.), *Physics and chemistry of the Earth*. Pergamon Press, Oxford, 940pp.

Hazen, R.M. (2005). Genesis: Rocks, minerals and the geochemical origin of life. *Elements*, 1, 135–38.

Healy, J.F. (1986). Pliny on mineralogy and metals, 111–46. In R. French and F. Greenaway (Eds.), *Science in the Early Roman Empire*. Croom Helm Ltd, London, 287pp.

Hewitt, J. (1922). On the zoological evidence relating to ancient land connection between Africa and other portions of the southern hemisphere. *Twentieth Annual Report of the South African Association for the Advancement of Science*, 316–31.

Holmes, A. (1927). *The age of the Earth*. Ernest Benn Ltd, 80pp.

—. (1965). *Principles of Physical Geology*. Thomas Nelson, London, 1288pp.

Hoover, H.C. and L.H. Hoover (1912, reprinted 1950). *Translation of De Re Metallica by Georgius Agricola (1556)*. Dover Publications, New York, 638pp.

Huffman, T.N. (1996). *Snakes and crocodiles: Power and symbolism in ancient Zimbabwe*. Witwatersrand University Press, Johannesburg, 228pp.

Huntress, W.T. (2003). Director's Report, Geophysical Laboratory. *Yearbook, Carnegie Institute of Washington*, 2002–2003.

Huppert, H.E. and J.S. Turner (1981). Double diffusive convection. *Journal of Fluid Mechanics*, 106, 299–329.

Huthnance, J. (2003). Sea change. *Geoscientist*, Geological Society of London, 13: 6, 4–7.

Irvine, T.N. and C.H. Smith (1967). Ultramafic rocks of the Muskox intrusion, Northwest Territories, Canada, 38–49. In P.J. Wyllie (Ed.), *Ultramafic and related rocks*. John Wiley and Sons, New York, 464pp.

Irvine, T.N., D.W. Keith and S.G. Todd (1983). The J-M platinum-palladium reef of the Stillwater complex, Montana. *Economic Geology*, 78, 1287–334.

Jackson, E.D. (1971). Origin of ultramafic rocks by cumulus processes. *Fortschrifte der Mineralogie*, 48, 128–74.

Jeffreys, Sir H. (1929). *The Earth*. Cambridge University Press, Cambridge, 346pp.

Johnson, W. (1909). The origin and formation of the diamond. *Seventh Annual Report of the South African Association for the Advancement of Science*, 167–80.

—. (1914). The origin and formation of the diamond. *Twelfth Annual Report of the South African Association for the Advancement of Science*, 275–86.

Jones, R.T. (1999). Platinum smelting in South Africa. *South African Journal of Science*, 95, 525–34.

Kanfer, S. (1993). *The last Empire*. Hodder and Stoughton, London, 409pp.

Kermode, F. (1985). Apocalypse and the Modern, 84–106. In S. Friedländer, G. Holton, L. Marx and E. Skolnikoff (Eds.), *Visions of Apocalypse: End or rebirth?* Holmes and Meier Publishers Ltd, London, 268pp.

King, L.C. (1962). *The morphology of the Earth*. Oliver and Boyd, London, 699pp.

Kirk, J., J. Ruiz, J. Chesley, J. Walshe and G. England (2002). A major Archaean, gold- and crust-forming event in the Kaapvaal craton, South Africa. *Science*, 297, 1856–58.

Kirkley, M.B., J.J. Gurney and A.A. Levinson (1991). Age, origin and emplacement of diamonds. *Gems and Gemology*, 27, 2–25.

Kukla, G. (2004). Central Arctic: Battleground of natural and man-made climate forcing. *EOS, Transactions, American Geophysical Union*, 85, 20, 200.

Leakey, R. and R. Lewin (1992). *Origins reconsidered*. Abacus: Little, Brown and Co, London, 375pp.

Levin, G. (1991). The Pretoria Saltpan. *Geobulletin*, Geological Society of South Africa, Johannesburg, 34, 13–16.

Lloyd, P.J.D. (1978). Maximizing the recovery of gold from Witwatersrand ores. *Minerals Science and Engineering*, 10, 208–21.

Lombaard, A.F. and the O'okiep Exploration Department Staff (1986). The copper deposits of the Okiep District, Namaqualand, 1421–45. In C.R. Anhaeusser and S. Maske (Eds.), *Mineral deposits of Southern Africa*. Geological Society of South Africa, Johannesburg, 2334pp.

Lombaard, B.V. (1949). Discussion on paper by S. van Biljon. *Transactions of the Geological Society of South Africa*, 52, 181–86.

Long, D.G., J. Ballantyne and C. Bertoia (2002). Is the number of Antarctic icebergs really increasing? *EOS, Transactions, American Geophysical Union*, 83, 42, 469–74.

Lowe, D.R. and G.R. Byerly (2003). Ironstone pods in the Archean Barberton greenstone belt, South Africa. *Geology*, 31, 909–12.

Lyddeker, R. (1898). The Tertiary connection of South America with South Africa. *Transactions of the Geological Society of South Africa*, 3, 151–53.

Lyell, Sir Charles (1834). *Principles of geology*. John Murray, London, 420pp.

—. (1850). *Principles of geology*. John Murray, London, 811pp.

—. (1865). *Elements of geology*. John Murray, London, 794pp.

Maaloe, S. (1978). The origin of rhythmic layering. *Mineralogical Magazine*, 42, 337–45.

Macgregor, A.M. (1927). The problem of the Precambrian atmosphere. *South African Journal of Science*, 24, 155–72.

Marais, J.A.H., A.L.D. Agenbacht, M. Prinsloo and W.A. Basson (2001). *The geology of the Springbok area: Explanation of Sheet 2916*. Council for Geoscience, Pretoria, 103pp.

Mather, K.F and S.L. Mason (1939). *A source book in geology*. McGraw-Hill Book Co., New York, 702pp.

Maury, C.J. (1913). The bearing of recent discoveries of early Tertiary shells near Trinidad. *Eleventh Annual Report of the South African Association for the Advancement of Science*, 92–96.

McAlester, A. L. (1968). *The history of life*. Prentice-Hall, New Jersey, 152pp.

McBirney, A.R. and R.M. Noyes (1979). Crystallization and layering of the Skaergaard intrusion. *Journal of Petrology*, 20, 487–564.

McCarthy, T. and B. Rubidge (2005). *The story of Earth and life*. Struik, Cape Town, 335pp.

McCulloch, J. (2002). *Asbestos blues*. James Currey, Oxford, 223pp.

McDonald, D. (1960). *A history of platinum*. Johnson Matthey and Co, Ltd, London, 254pp.

McGowan, C. (2003). *The dragon seekers*. Abacus, London, 270pp.

Mellor, E.T. (1922). The influence of mineral deposits on the development of a young country. *Twentieth Annual Meeting of the South African Association for the Advancement of Science*, 42–51.

Mellor, E.T. and T.N. Leslie (1906). On a fossil forest recently exposed in the bed of the Vaal River at Vereeniging. *Transactions of the Geological Society of South Africa*, 9, 125–28.

Menard, H.W. (1986). *The ocean of truth*. Princeton University Press, New Jersey, 353pp.

Mennell, F.P. (1904). The prehistoric monuments of Rhodesia. *Second Annual Report of the South African Association for the Advancement of Science*, 509–18.

—. (1908). Notes on some diamond-bearing and associated rocks. *Sixth Annual Report of the South African Association for the Advancement of Science*, 105–06.

Merensky, H. (1907). The origin of river diamonds within the area of the Vaal. *Transactions of the Geological Society of South Africa*, 10, 107–11.

Metcalfe, Sir Charles (1904). Presidential address. *Second Annual Report of the South African Association for the Advancement of Science*, 15–30.

Metrowich, F.C. (1953). *Assegaai over the Hills*. Howard Timmins, Cape Town, 214pp.

Milkov, A.V., G.E. Claypool, Y-J. Lee, W. Xu, G.R. Dickens and W.S. Borowski (2003). In situ methane concentrations at Hydrate Ridge, offshore Oregon. *Geology*, 31, 833–36.

Miller, A.I. (1998). Biotic transitions in global marine diversity. *Science*, 281, 1157–60.

Miller, A.I. and M. Foote (2003). Increased longevities of post-Palaeozoic marine genera after mass extinctions. *Science*, 302, 1030–32.

Miller, D. (1995). 2000 years of mining. *South African Journal of Geology*, 98, 232–38.

Milton, J. (1674). *Paradise Lost*, 181–448. In H.C. Beeching (1925, Ed.), *The poetical works of John Milton*. Oxford University Press, Oxford, 554pp.

Minerals Bureau (2000). *South Africa's mineral industry, 1998–99*. Mineral Economics Directorate, Pretoria, 266pp.

Mintek Bulletin (1991). *Diamonds are (not) forever*. Information and Communications Division, Mintek, Randburg, No.42.

Molengraaff, G.A.F. (1897a). Diamonds at Rietfontein. *Transactions of the Geological Society of South Africa*, 3, 122–23.

—. (1897b). Report of the State Geologist of the S.A. Republic. *Transactions of the Geological Society of South Africa*, 4, 119–47.

—. (1898). The glacial origin of the Dwyka Conglomerate. *Transactions of the Geological Society of South Africa*, 4, 103–15.

—. (1903). Vice-Presidential address for 1902. *Transactions of the Geological Society of South Africa*, 5, 69–75.

—. (1904). *The geology of the Transvaal*. T. and A. Constable, Edinburgh, 90pp.

Molyneux, A.J.C. (1911). Some problems presented by the Karroo System in Rhodesia. *Ninth Annual Report of the South African Association for the Advancement of Science*, 31–45.

Moore, A.E. (1999). A reappraisal of epeirogenic flexure in southern Africa. *South African Journal of Geology*, 102, 363–76.

Moore, J.M. and A.E. Moore (2004). The roles of primary kimberlitic and secondary glacial sources. *Journal of African Earth Sciences*, 38, 115–34.

Moore, J.R. (1973). Charles Lyell and the Noachian deluge, 405–25. In A. Dundes (1988, Ed.), *The Flood myth*. University of California Press, Berkeley, 452pp.

Moore, R. (1954). *Man, time and fossils*. Jonathan Cape, London, 382pp.

Morrison, J.T. (1918). On the internal structure of the earth. *Sixteenth Annual Report of the South African Association for the Advancement of Science*, 31–44.

Narbonne, G.M. and J.G. Gehling (2003). Life after snowball. *Geology*, 31, 87–90.

Neuse, E.W. (1999). Platinum coordination compounds in cancer research and chemotherapy. *South African Journal of Science*, 95, 509–16.

Nisbet, E.G. and H.N. Sleep (2001). The habitat and nature of early life. *Nature*, 409, 1083–91.

Nutman, A.P., C.R.L. Friend and V.C. Bennett (2001). Review of the oldest (4400–3600Ma) geological and mineralogical record. *Episodes: Journal of International Geoscience*, 24, 93–101.

Okafor, V.I. and N.J. Coville (1999). Platinum in catalysis. *South African Journal of Science*, 95, 503–08.

Olsen, P. (2002). The disappearing dipole. *Nature*, 416, 591–94.

Orlov, Y.A. (1962). *Fundamentals of Palaeontology*. Israel Programme for Scientific Translations, Jerusalem, 482pp.

Pakenham, T. (1982). *The Boer War*. Futura Publications, London, 659pp.

Partridge, T.C., W.U. Reimold, S.J. Kerr and I.G. Stanistreet (1991). The Pretoria Saltpan. *Geobulletin of the Geological Society of South Africa*, 34, 16–22.

Pedersen, O. (1986). Some astronomical topics in Pliny, 162–96. In R. French and F. Greenaway (Eds.), *Science in the Early Roman Empire*. Croom Helm Ltd, London, 287pp.

Penning, W.H. (1888). The South African goldfields. *Journal of the Society of Arts (London)*, 36, 433–44.

Peters, C. (1902). *The Eldorado of the Ancients*. Reprint edition by Books of Rhodesia Publishing Co., 1977, 447pp.

Pettijohn, F.J. (1957). *Sedimentary rocks*. Harper and Brothers, New York, 718pp.

Pikirayi, I. (2001). *The Zimbabwe culture*. Alta Mira Press, Walnut Creek, Calif., 303pp.

Popov, Y.A., S.L. Pevzner, V.P. Pimenov and R.A. Romushkevich (1999). New geothermal data from the Kola superdeep well SG-3. *Tectonophysics*, 306, 345–66.

Prister, A. (1898a). The glacial origin of the Dwyka Conglomerate. *Transactions of the Geological Society of South Africa*, 4, 151–52.

—. (1898b). The origin and formation of the Witwatersrand auriferous conglomerates. *Transactions of the Geological Society of South Africa*, 4, 19–36.

Qixiang, H. (2001). A review of the catastrophic extinction at the end of the Cretaceous, 361–72. In U. Briegel and W. Xiao (Eds.), *Paradoxes in Geology*. Elsevier Science B.V., Amsterdam, 467pp.

Randles, W.G.L. (1975). *The Empire of Monomotapa*. English translation reprint (1981), Mambo Press, Salisbury, Zimbabwe, 149pp.

Recknagel, R. (1908). Mineral deposits in the Rooiberg district. *Transactions of the Geological Society of South Africa,* 11, 83–106.

Reiche, H.A.T. (1985). The archaic heritage, 21–43. In S. Friedländer, G. Holton, L. Marx and E. Skolnikoff (Eds.), *Visions of Apocalypse: End or rebirth?* Holmes and Meier Publishers Ltd, London, 268pp.

Reimold, W.U., D. Brandt, R. de Jong and J. Hancox (1999). The Tswaing meteorite crater. *Popular Geoscience Series*, 1, Council for Geoscience, Pretoria, 177pp.

Reitz, D. (1943). *No Outspan*. Faber and Faber, Ltd, London, 288pp.

Repcheck, J. (2004). *The man who found time – James Hutton*. Perseus Books, Cambridge MA, 247pp.

Rice, C.M. (1956). *Dictionary of geological terms*. Princeton, New Jersey, 645pp.

Richardson, S.H. (1986). Latter-day origin of diamonds of eclogitic paragenesis. *Nature*, 322, 623–26.

Richardson, S.H., A.J. Erlank, J.W. Harris and S.R. Hart (1990). Eclogitic diamonds of Proterozoic age from Cretaceous kimberlites. *Nature*, 346, 54–56.

Richardson, S.H., J.J. Gurney, A.J. Erlank and J.W. Harris (1984). Origin of diamonds in old enriched mantle. *Nature*, 310, 198–202.

Ringwood, A.E. (1975). *Composition and petrology of the Earth's mantle*. McGraw-Hill Book Co., New York, 618pp.

Roberts, D.L. (1998). Mid- to Late Pleistocene fossil human footprints in coastal aeolianites, South Africa. *Conference of the International Association for the Study of Human Biology*. Sun City, South Africa.

Robertson, M. (1974). *Diamond fever: South African diamond history*. Oxford University Press, London, 250pp.

Rogers, A.W. (1917). An old report on the copper field of Namaqualand. *Fifteenth Annual Report of the South African Association for the Advancement of Science*, 146–52.

Rogers, A.W., A.L. Hall, P.A. Wagner and S.H. Haughton (1929). *The geology of the Union of South Africa.* Carl Winters Universitätsbuchhandlung, Heidelberg, 232pp.

Romanovsky, V., S. Smith, K. Yoshikawa and J. Brown (2002). Permafrost temperature records. *EOS, Transactions, American Geophysical Union*, 83, 589–94.

Rona, P.A., D.R. Jackson, K.G. Bemis, C.D. Jones, K. Mitsuzawa, D.R. Palmer and D. Silver (2002). Acoustics advances study of sea floor hydrothermal flow. *EOS, Transactions, American Geophysical Union*, 83, 497–502.

Rosenthal, E. (1950). *Shovel and sieve.* George Allen and Unwin Ltd, London, 199pp.

Roseveare, W.N. (1917). Mathematical analysis and science. *Fifteenth Annual Report of the South African Association for the Advancement of Science*, 33–47.

Runcorn, S.K. (1959). Rock magnetism, 271–90. In J.F. White (Ed.), *Study of the Earth.* Prentice-Hall, New Jersey, 408pp.

Ruskin, J. (1880). *The seven lamps of architecture.* George Allen, Kent, 222pp.

Ryan, W. and W. Pitman (2000). *Noah's Flood.* Touchstone, Simon and Schuster, London, 337pp.

Sampson, E. (1932). Magmatic chromite deposits in Southern Africa. *Economic Geology*, 27, 113–44.

Sandberg, C.G.S. (1928). The origin of the Dwyka conglomerate of South Africa. *Geological Magazine*, London, 65, 117–39.

Schenck, A. (1895). On glacial phenomena in South Africa. *Transactions of the Geological Society of South Africa*, 1, 129–41.

Schewe, P.F. (2003). Physics update. *Physics Today*, 56, 9.

Schiermeier, Q. (2004). Noah's flood. *Nature*, 430, 718–19.

Schwarz, E.H.L. (1908). Discoveries of economic importance made by the Albany pioneers. *Sixth Annual Report of the South African Association for the Advancement of Science*, 49–63.

—. (1909). The bearing of recent theories on the nature of the Earth's interior. *Seventh Annual Report of the South African Association for the Advancement of Science*, 234–41.

—. (1912). *South African geology.* Blackie and Son Ltd, London, 200pp.

—. (1914a). The Kimberley mines. *Twelfth Annual Report of the South African Association for the Advancement of Science*, 314–20.

—. (1914b). The lost land of Agulhas. *Twelfth Annual Report of the South African Association for the Advancement of Science*, 169–79.

—. (1920). *The Kalahari.* Maskew Miller, Cape Town, 163pp.

—. (1923). The Kalahari Scheme. *Twenty-first Annual Report of the South African Association for the Advancement of Science*, 208–22.

Showstack, R. (2002). Antarctic 'Larsen B' ice shelf shatters. *EOS, Transactions, American Geophysical Union*, 83, 14, 158.

Skinner, B.J. and S.C. Porter (1999). *The dynamic Earth*. John Wiley and Sons, Inc., New York, 575pp.

Smalberger, J.M. (1975). *A history of copper mining in Namaqualand*. C. Struik Ltd, Cape Town, 152pp.

Smith, G. (1873). The Chaldaean account of the Deluge, 29–48. In A. Dundes (1988, Ed.), *The Flood myth*. University of California Press, Berkeley, 452pp.

Smith, G.A. (1972). Continental drift, 213–31. In I.G. Gass, P.J. Smith and R.C.L. Wilson, (Eds.), *Understanding the Earth*. Artemis Press, Sussex, 383pp.

Smith, G.F.H. (1940). *Gemstones*. Methuen and Co., London, 443pp.

Smith, J.V. (2005). Geochemical influences on life's origins and evolution. *Elements*, 1, 151–56.

Smith, R. (1995). Life in a prehistoric crater-lake. *Phoenix Magazine*. Albany Museum, Grahamstown, 8, 2 and 3, 4–6.

Snyder, D.B. and R.W. Hobbs (1999). Deep seismic reflection profiles across the Chicxulub crater, 263–68. In B.O. Dressler and V.L. Sharpton (Eds.), *Large meteorite impacts and planetary evolution*. Geological Society of America Special Paper, 339, 464pp.

Sobel, D. (1999). *Galileo's daughter*. Harper Collins Audio Books, London.

Sollas, W.J. (1905). *The age of the Earth*. T. Fisher Unwin, London, 328pp.

Somoza, L., J.M. Gardner, V. Diaz-del-Rio, J.T. Vazquez, L.M. Pinheiro and F.J. Hernandez-Molina (2002). Numerous methane gas-related sea floor structures in Gulf of Cadiz. *EOS, Transactions, American Geophysical Union*, 83, 541–49.

Söhnge, P.G. (1946). *Geology of the Messina copper mines*. Memoir 40, Geological Survey of South Africa, Pretoria, 277pp.

Spamer, C. (1996). Of myths and middens: The story of Thulamela. *Gold Fields of South Africa Annual Review for 1995–1996*, 4–17.

Stanley, S.M. and M.G. Powell (2003). Depressed rates of origination and extinction during the Late Palaeozoic Ice Age. *Geology*, 31, 877–80.

Stirling, J. (Ed.). (1954). *The Bible, authorised version*. British and Foreign Bible Society, London, 256pp.

Stow, G.W. (1870). On some points in South African geology. *Proceedings of the Geological Society of London*, 27, 497–547.

Suess, E. (1904). *The face of the Earth*. Clarendon Press, Oxford, 604pp.

Summers, R. (1969). *Ancient mining in Rhodesia*. Memoir 3, National Museums of Rhodesia, 236pp.

Sylvester-Bradley, P.C. (1972). An evolutionary model for the origin of life, 123–41. In I.G. Gass, P.J. Smith and R.C.L. Wilson (Eds.), *Understanding the Earth*. Artemis Press, Sussex, 383pp.

Tankard, A.J., M.P.A. Jackson, K.A. Eriksson, D.K. Hobday, D.R. Hunter and W.E.L. Minter (1982). *Crustal evolution of Southern Africa.* Springer-Verlag, New York, 523pp.

Tarbuck, E.J. and F.K. Lutgens (1984). *The Earth.* Charles Merrill Publishing Company, Toronto, 594pp.

Taylor, S.S. and T. Biggs (1999). Innovations in platinum jewellery materials. *South African Journal of Science*, 95, 543–49.

Thiel, S. (1906). Report to General Meeting of Shareholders. *The Prince Albert Goldfields Prospecting and Proprietary Syndicate*, 1p.

Torrente, M. (1904). Evolution of the treatment of by-products on the Witwatersrand goldfields. *Second Annual Report of the South African Association for the Advancement of Science*, 2, 74–82.

Trevor, T.T. and E.T. Mellor (1908). Report on a reconnaissance of the northwestern Zoutpansberg district. *Report of the Transvaal Chamber of Mines*, Johannesburg.

Truter, F.C. (1949). Discussion on paper by S. van Biljon. *Transactions of the Geological Society of South Africa*, 52, 178–81.

Umbgrove, J.H.F. (1947). *The pulse of the Earth.* Martinus Nijhoff, the Hague, 358pp.

Unattributed (1993). A mineral as recognised by law. *Geobulletin of the Geological Society of South Africa*, 36, 38.

Van Biljon, S. (1949). The transformation of the upper part of the Pretoria Series in the Bushveld Igneous Complex. *Transactions of the Geological Society of South Africa*, 53, 1–174.

—. (1958). Contributed discussion on paper by F.J. Coertze. *Transactions of the Geological Society of South Africa*, 61, 394–97.

Van Wyk, T. (1995). The splendour of ancient coins. *Gold Fields of South Africa Annual Review for 1994–1995*, 34–39.

Vermaak, C.F. and M.J. van der Merwe (2000). *The platinum mines and deposits of the Bushveld Complex, South Africa.* Mintek, Randburg, 118pp.

Viljoen, M.J. (1999). The nature and origin of the Merensky Reef. *South African Journal of Geology*, 102, 221–39.

Voit, F.W. (1907a). Further remarks on the kimberlite rock and the origin of diamonds. *Transactions of the Geological Society of South Africa*, 10, 101–06.

—. (1907b). The origin of diamonds. *Transactions of the Geological Society of South Africa*, 10, 75–80.

Von Gruenewaldt, G. (1970). Contributed discussion on paper by Hamilton (1970), 376–77. In D.J.L. Visser and G. von Gruenewaldt (Eds.), *Symposium on the Bushveld Igneous Complex and other layered intrusions.* Special Publication of the Geological Society of South Africa, Johannesburg, 1, 763pp.

Wager, L.R. and G.M. Brown (1968). *Layered igneous rocks*. Oliver and Boyd Ltd, London, 588pp.

Wagner, P.A. (1914). *The Diamond Fields of South Africa*. The Transvaal Leader, Johannesburg, 347pp.

—. (1917). Some problems in South African geology. *Proceedings of the Geological Society of South Africa*, 20, xix–xxiv.

—. (1922). *The Pretoria Saltpan, a soda caldera*. Memoir 20, Geological Survey of South Africa, Pretoria, 136pp.

Wannenburgh, A.J. and P. Johnson (1990). *Diamond people*. Norfolk House Publishers, London, 280pp.

Webby, B.D., M.L. Droser and F. Paris (2004). Final Report of IGCP Project 410: The Great Ordovician biodiversification event. *Episodes, Journal of International Geoscience*, 27, 203–08.

Wegener, A. (1929). *The origin of continents and oceans* (Translated from the Fourth revised German edition by John Biram [1966]). Methuen and Co., London, 248pp.

Werdmüller, V.W. (1986). The Central Rand, 7–47. In E.S.A. Antrobus (Ed.), *Witwatersrand gold – 100 years*. Geological Society of South Africa, Johannesburg, 298pp.

Westall, F., M.J. de Wit, J. Dann, S. van der Gaast, C.E.J. de Ronde and D. Gerneke (2001). Early Archaean fossil bacteria and biofilms. *Precambrian Research*, 106, 93–116.

Wezel, F.C. (2005). IGC 2004 – Urbino international workshop on global tectonics. *Episodes, Journal of International Geoscience*, 28–1, 55–57.

White, F. (1903a). Dry crushing of ore preparatory to the extraction of gold. *First Annual Report of the South African Association for the Advancement of Science*, 362–68.

—. (1903b). The less known ruins in Rhodesia. *First Annual Report of the South African Association for the Advancement of Science*, 480–91.

White, J.F. (1962). *Study of the Earth*. Prentice-Hall, New Jersey, 408pp.

Wilde, S.A., J.W. Valley, W.H. Peck and C.M. Graham (2001). Evidence from detrital zircons for the existence of continental crust and oceans on the Earth 4.4 Gyr ago. *Nature*, 409, 175–78.

Williams, A.F. (1932). *The genesis of the diamond, Vols.1–2*. Ernest Benn Ltd, London, 640pp.

Williams, G.F. (1906). Presidential address. *Fourth Annual Report of the South African Association for the Advancement of Science*, 3, 1–10.

Winchester, S. (2001). *The map that changed the World*. Viking Press, Penguin Group, London, 338pp.

Wolff, I.M. (1999). New applications for ruthenium. *South African Journal of Science*, 95, 539–43.

Woolley, L. (1956). Stories of the Creation and the Flood, 89–99. In A. Dundes (1988, Ed.), *The Flood myth*. University of California Press, Berkeley, 452pp.

Worst, B. (1991). The discovery of platinum in South Africa. *Quarterly News Bulletin of the Geological Society of South Africa*, 34, 20–21.

Wyllie, P.J. (1971). *The dynamic Earth*. John Wiley and Sons, New York, 416pp.

Yin, C.Y., F. Tang, Y. Liu, L. Gao, P. Liu, Y. Xing, Z. Yang, Y. Wan and Z. Wang (2005). U-Pb zircon age from the base of the Ediacaran Doushantuo Formation in the Yangtze Gorges, South China. *Episodes, Journal of International Geoscience*, 28, 48–49.

Young, R.B. (1909). Further notes on the auriferous conglomerates of the Witwatersrand. *Transactions of the Geological Society of South Africa*, 12, 82–101.

—. (1911). Presidential address: The problem of the Rand banket. *Proceedings of the Geological Society of South Africa*, 14, xxi–xxx.

—. (1913). Presidential address. *Eleventh Annual Report of the South African Association for the Advancement of Science*, 17–23.

—. (1917). *The banket*. Gurney and Jackson, London, 125pp.

Young, R.B. and J.P. Johnson (1906). Glacial phenomena in Griqualand West. *Transactions of the Geological Society of South Africa*, 9, 34–39.

Index